T0310094

Polyurethane Immobilization of Cells and Biomolecules

Polyurethane Immobilization of Cells and Biomolecules

Medical and Environmental Applications

T. Thomson

This edition first published 2018

Registered Office
John Wiley & Sons, Inc., 111 River Street, Hoboken, NJ 07030, USA

Editorial Office
111 River Street, Hoboken, NJ 07030, USA

For details of our global editorial offices, customer services, and more information about Wiley products visit us at www.wiley.com.

Wiley also publishes its books in a variety of electronic formats and by print-on-demand. Some content that appears in standard print versions of this book may not be available in other formats.

Library of Congress Cataloging-in-Publication Data

Names: Thomson, T. (Tim), author.
Title: Polyurethane immobilization of cells and biomolecules : medical and environmental applications / by T. Thomson.
Description: Hoboken, NJ : John Wiley & Sons, 2018. | Includes bibliographical references and index. |
Identifiers: LCCN 2017036434 (print) | LCCN 2017044756 (ebook) | ISBN 9781119264941 (pdf) | ISBN 9781119264965 (epub) | ISBN 9781119254690 (cloth)
Subjects: LCSH: Immobilized cells. | Polyurethanes–Biotechnology. | Polyurethanes–Industrial applications.
Classification: LCC TP248.25.I55 (ebook) | LCC TP248.25.I55 T56 2018 (print) | DDC 668.4/239–dc23
LC record available at https://lccn.loc.gov/2017036434

Cover design by Wiley
Cover image: Copyright and courtesy of Linda Bradford

Set in 10/12pt Warnock by SPi Global, Pondicherry, India

Printed in the United States of America

10 9 8 7 6 5 4 3 2 1

Contents

Preface

In the next several hundred pages, we will be describing a virtual laboratory. The lab has three sections:

- Scaffold development
- Immobilization technologies
- Controlled release

Each section will be guided by research on the treatment of recalcitrant pollutants or the development of organ assist devices, specifically the liver and the pancreas. Describing the objectives, plans, and goals of each section is the purpose of this book. Whether the goal is environmental or medical science relater, each section is benefited by the research in the others. We will use research from around the world to show how those concepts can be built into polyurethane chemistry. We will show how the goals of each section are met by a short list of raw materials:

1) A commercially reticulated polyurethane foam
2) Polyethylene glycol (1000, 4000, and 10 000 molecular weights)
3) Toluene diisocyanate
4) Trimethylolpropane

This is not a chemistry book, however. It is the application of chemistry to two of our most important technical challenges, specifically the remediation of polluted air and water and the development of hybrid artificial organs. The later is to meet a permanent shortage of transplantable body parts. While we recognize that these are as different from one another as they can be, we will make the case that the technology to solve one problem is the technology that can solve the other. Consider the human liver. It is a flow-through device that, among other things, metabolizes components in blood passing through it. Compare this to a tank or column that is packed with a medium to which bacterial cells or enzymes have been immobilized. It is a flow-through device that metabolizes components in a fluid passing through it. In both cases, there are minimum requirements for the device to function. Among these are permeability and surface area to permit an efficient conversion. These will be explained in detail.

As such, this book is directed toward biotechnologists, specifically whether they are environmental engineers or medical researchers. Having said that, polymer chemists will find it as useful as a comprehensive discussion of a leading edge of polymer technology. Those in the polyurethane industry will see it as a useful extension of this unique polymer chemistry. We will make the case that polyurethane is an ideal chemistry to approach these challenging applications. We will also make the case, probably till you are bored hearing about it, that polyurethane is not a molecule but rather a system composed of several parts, each of which adds to the resultant polymer. For example, it can be hydrophilic or hydrophobic or somewhere in between. It is what we call *amicas hydrophilii*. Small changes in chemistry allow it to be used as a wound dressing or an automobile fender. The physical forms that polyurethane can take are equally diverse. It can be an elastomer (e.g., for an automobile fender) or a bridge support component. In your local drug store, you can find cosmetic applicator sponges made from polyurethane. Most remarkably, it can be processed such that it is almost not there. Polyurethane sponges can be made with a void volume of 97%. During processing a small amount of water in the formulation changes the resultant polymer from a foam to a hard polymer to an adhesive. We will talk about flow-through and surface area. These sponges have virtually no resistance to fluids passing through it and with surface areas approaching 7000 m^2/m^3. The result is a large surface that can be used for a number of applications without inhibiting the flow of fluids. We will be exploring these concepts in detail.

We will describe research done in and for our labs and the research of others in the use of polyurethane and other chemistries as an immobilizing agent for cells and what we call active molecules. Cells include organisms from bacteria to mammalian cells. Active molecules include not only enzymes but also, as we will discuss, cell attachment and other ligands. As we said applications range from not only environmental remediation to clinical but also analytical and diagnostic techniques. We will use the term architecture many times. In the sense of this text, architecture represents a three-dimensional structure. Not to jump too far ahead of ourselves, but the human liver has a recognizable shape. This is the result of not only cell–cell communication among the cells but also the scaffold within and on which the organ develops.

In probably the most important chapters of the book, we will describe how specific architectures of polyurethanes are made and are then used to support living cells for medical and environmental applications. This identifies the material as a scaffold. That is to say there are many applications for which polyurethanes are used, but when the application is for the support of living cells or biomolecule, we refer to it as a scaffold. This allows us to focus on the applications that are the subject of this book as opposed to the thousands of uses for this unique polymer system.

For the biotechnologists, let us warn you that we are chemists. What we know of the subject we will be discussing is based on work we have done with professionals and from the literature. We have sponsored research at various labs and universities, and although we cannot call ourselves expert, we are confident that the technology herein described is real and valuable.

To begin the discussion, it is necessary to describe chemistry. Don't be concerned. While the discussion is comprehensive, it is not complicated. The first chapter is a graduate-level course in polyurethane but only requires introductory knowledge of general chemistry. As we will discuss, polyurethanes have several parts, each of which influence the characteristics of the resultant polymer. At the end of the chemistry chapter, you will begin to know what parts might meet your individual requirements. Then the information in the chapters on controlled release and immobilization will complete your education.

Having said that, there are several companies that make the raw materials for your research. Therefore, while your research might eventually design your own polymers, it is convenient to begin with commercial materials. As you develop skills in the techniques, and even develop novel techniques, you may have a need to make adjustments in the basic chemistry. For example, you may need a stiffer material or more flexible. Polyurethanes offer a convenient way of making those changes. More appropriately, we will be discussing biodegradability and biocompatibility, both of which are far from being resolved. Regardless of your training we would advise you to go through chemistry in order to see the context with the rest of the book.

By way of introduction, we were part of the hydrophilic polyurethane (HPUR) commercial venture at the W. R. Grace Corp. The trade name for the family of products was Hypol™ prepolymer, still the dominant producer of HPUR products. I was assigned to support the existing sales base and expand the applications. In the several years I spent in that position, I had the pleasure to travel the world explaining the benefits of this unique chemistry. The product markets ranged from personal care products to advanced medical devices to agriculture. After leaving Grace, I organized *Main Street Technologies* as a venue for my personal research interests, writing several books, and limited consulting. During that period we took several assignments in manufacturing units. This expanded my knowledge of polyurethanes with day-to-day experience in the manufacture of foam. We always maintained a research focus, however.

While the metaphor of "standing on the shoulders of giants" is commonly used, I refer to my career as that of a student. The men I have worked with and for, and the customers that I tried to help, have been my teachers. I have taken what has been taught to me and applied it to my own research. I can only hope that I have earned a passing grade. In any case, this book is in part dedicated to them. More important than that I dedicate this book to my wife, Maguy. Her love helped me from a wild eye kid to something resembling a scientist.

This book is unique in a sense in that it speaks to two audiences, typically considered sufficiently different to be considered other sciences. We work in both areas without confusion, but in an effort to speak to both audiences simultaneously, we must rigorously avoid jargon. Those of you who have tried to be technical generalists will understand the difficulty in walking that line. As an example of what we need to avoid, consider the following:

> "this spiral arrangement of collagen fibers with their adjacent smooth muscle cell layer allows the small intestines to constrict in a manner that promotes the efficient transport of a bolus of biomass."

Most of us know this process by other names.

Lastly, when you as an environmentalist read the sections on medical research, when they say blood, mentally transpose that into air or water. It will make perfect sense. Conversely, as medical professionals, when reading about environmental issues, replace references to air and water to blood. You will see the continuity.

Cover Art

We were asked by a New York artist to help her find a replacement for a brush that she had used to create the effect seen on the cover. For whatever reason, she was not able to find replacement brushes, and so she was not able to duplicate her innovative technique. To make a long story short, we determined that the effect was due to a number of factors. Pore structure, size, and architecture, which control the flow, were the most important. We also found that surface chemistry (wetting) and chemistry of the paints were critical.

As you go through this text, you will see that these sane properties will be mentioned over and over again as we develop our arguments. We, therefore, thought it would be appropriate.

1

Polyurethane Chemistry

Introduction

There are many texts on polyurethanes (PURs) but this one has a special interest. After the first couple of chapters, we will focus on how this chemistry can be used to advance the sciences of environmental remediation and medical science. While those may seem too diverse for a single volume, we think we can make the case that there is a unifying aspect, and, furthermore, it is PURs that best fit that role. Polyurethanes are remarkable in the world of polymers in that they are not a molecule like polyethylene or polyvinylchloride, but rather a system with multiple component parts. Each of those parts fulfills a certain and individual function. It is their selection and the methods used to process the polymer that make it unique. With the help of this book, a scientist with ordinary knowledge of chemistry can learn these techniques. Furthermore, unlike the more common polymers, innovative research can be developed in the average laboratory setting. Among other things, you will learn how to make products from elastomers to foams to adhesives with only slight changes in chemistry or processes. Applying those simple skills with the experience taught in the final chapters, the reader is offered the potential to conduct world-class research in fields from water and air treatment to artificial organs. A bold claim, but defendable.

To begin, PURs are a family of polymers all based on the reaction of an organic isocyanate and a multifunctional polymer. Isocyanates, as we will discuss, react quickly with other compounds like water, amines, alcohols, and organic acids. The defining aspect of a PUR is the isocyanate starting material. Because of its somewhat unique reactivity, one can build a polymer of his or her own design. It is what you react the isocyanate with that defines the characteristic of the resultant PUR. For example, with the same isocyanate one can produce a hydrophobic or hydrophilic foam and a seat cushion or a dressing to

Polyurethane Immobilization of Cells and Biomolecules: Medical and Environmental Applications, First Edition. T. Thomson.

treat dermal ulcers. As this book develops we hope to illustrate the range of products and technologies that are possible with the knowledge taught in this chapter and the talents of the reader.

As we mentioned, PURs are a combination of several parts. We will describe each of these but a history lesson is appropriate. The first official PURs were developed prior to World War II. It was first produced as a replacement for natural rubber. Otto Bayer and his coworkers at I.G. Farben in Leverkusen, Germany, made PURs in 1937. The first PURs were hydrophilic. Their intended use was for automobile tires, but the polymers were not strong enough to withstand the weight of a car when wet. It wasn't until hydrophobic polyols were used that it became the useful material we know today. It was in the 1950s that Monsanto developed the so-called "one-step" process to make foam that made PURs economically viable in a wide range of product markets. The campaign to reduce weight and cost catalyzed the expansion of PUR elastomers in automobile parts. Currently, applications range from furniture foams to elastomers to adhesives for home and industrial use. We remind you that this has happened without major changes in the chemistry. In the 1970s a hydrophilic version was redeveloped and numerous unique applications researched, including the immobilization of biomolecules and cells. This research led to the international hydrophilic PUR industry. It is our opinion that this product and derivatives thereof will provide a path into expanded medical and environmental uses.

The Chemistry

Commercial PURs are the result of the exothermal reaction between an isocyanate and a molecule containing two or more alcohol groups (–OH). While this defines current commercial applications, the chemistry is not limited to alcohols, as we will explain. The properties of the resultant PUR depend on the choice of these components. If the application is as a consumer product, both cost and strength of materials guide the development and so appropriate components are selected. If the product is to be biocompatible or come into contact with blood, a different set of components will be necessary and cost may not be a critical factor.

In either case, Figure 1.1 shows the reaction of an isocyanate and an alcohol. The result illustrates the urethane linkage. One can imagine the polymerization using a diisocyanate and molecules with multiple –OH end groups.

$$R{-}N{=}C{=}O + R'{-}CH_2{-}OH \longrightarrow R{-}\underset{\displaystyle H}{\underset{|}{N}}{-}\overset{\displaystyle O}{\overset{||}{C}}{-}O{-}CH_2{-}R'$$

Figure 1.1 The urethane reaction.

There are many isocyanates and polyols to choose from and these are the tools of the trade to a urethane chemist. While we will see that there is a limited supply when it comes to the choices of isocyanate, there is no limit to possible reactants. We will explore this in detail when we focus particularly on medical products. In that discussion we will report on research that uses modified polypeptides as replacements for conventional polyols. For clarity, what "R" represents is the subject of much research around the world.

To investigate this further, we will look at the components in more detail.

The Isocyanates

The world of commercial PURs is predominantly split between two isocyanates: toluene diisocyanate (TDI) and methylene-bis-diphenyldiisocyanate (MDI). Both of these are considered "aromatic" as they are built around the benzene ring. This has product shelf-life implications (Figure 1.2).

Their relative importance depends on a number of factors. TDI was the first successful isocyanate and is still important. It is relatively inexpensive, and due in part to its molecular weight (MW), the properties of the PUR from which it is made are more sensitive to the polyol.

We will be using a convention when describing polymers of this type. The isocyanate portion of a polymer is said to be a "hard" segment due to its MW and inability of the molecule to rotate within itself. The polyol, however, is a longer molecule and has a high degree of internal rotation. It is, therefore, referred to as "soft." Thus a polymer with a higher mass percent of isocyanate would tend to be stiffer/harder, and vice versa.

Polymers made from TDI are generally softer because of the relative weights of isocyanate and polyol, which is the preferred isocyanate for hydrophilic PURs. The higher percentage of polyol makes for more hydrophilic foam as well.

The bulk of the conventional PUR business, however, has shifted toward MDI as the isocyanate of choice. MDI is sold in different forms. In any case, its higher MW means that it is a portion of the resultant polymer with a higher weight. This makes it "harder" and more hydrophobic. This has strong implications for product characteristics. There are hydrophilics based on MDI but they tend to make more "boardy" foams due, in part, again to its increased mass % in the urethane molecule.

Figure 1.2 The aromatic diisocyanates.

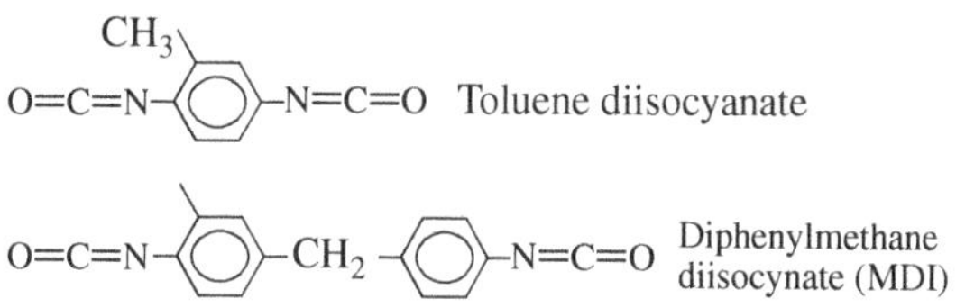

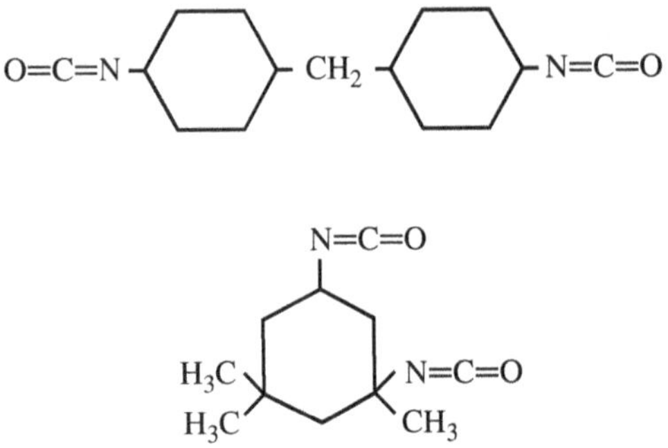

Figure 1.3 Aliphatic diisocyanates (top is hydrogenated MDI, below is isopherone diisocyanate).

While the so-called aromatics (TDI and MDI) represent the dominate isocyanates in the conventional and hydrophilic PUR businesses, they have a problem with respect to weathering, specifically yellowing on exposure to light and heat. While this may seem to be insignificant, the aesthetics of a product made from these materials is typically important. Whether the device is a cosmetic applicator or a wound dressing, yellowing is typically viewed as a degradation of the usefulness of the product. There is no evidence that the physical or hydrodynamic properties are affected by normal yellowing, but it is almost always an issue.

Three processes cause the yellowing. Exposure to UV light causes the production of color bodies in aromatic isocyanates (TDI, MDI, etc.). This can be inhibited by the use of UV-absorbing compounds. Most commonly, however, is to use packaging that is opaque to the ultraviolet.

Another major cause of yellowing is heat. Temperatures above 105°C can noticeably yellow foam in a few minutes. Ring opening and the resultant conjugated structures are thought to be the cause.

Lastly, exposure to hydrocarbon emissions causes yellowing. For this reason, hydrophilic PUR foam manufacturers typically use electric forklift trucks. As we will explain in the chapter on immobilization, PURs have a unique ability to absorb hydrocarbons from the air due to the polyol part of the molecule, which, again as we will discuss, is well known as a solvent extraction medium.

When the yellowing has to be eliminated (as opposed to inhibited), other isocyanates are available. The most common are the aliphatics shown in Figure 1.3.

You will notice that these compounds still have the six-member ring component, but, in this case, the ring is cyclohexane. It does not absorb UV of sufficient energy to produce the yellowing effect observed with TDI and MDI.

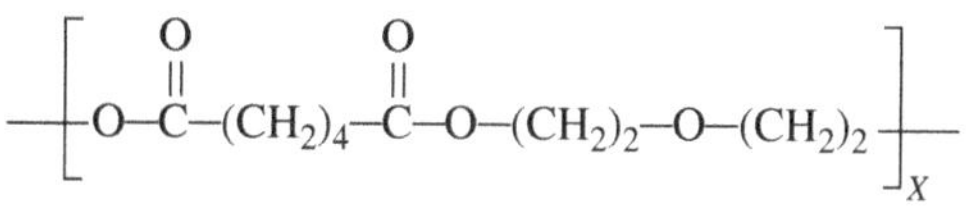

Figure 1.4 Polyester polyols.

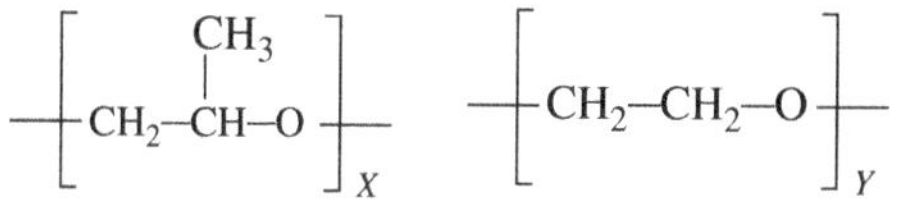

$$\left[-O-\overset{O}{\overset{\|}{C}}-(CH_2)_4-\overset{O}{\overset{\|}{C}}-O-(CH_2)_2-O-(CH_2)_2- \right]_X$$

Figure 1.5 Polyether polyols.

$$\left[-CH_2-\overset{CH_3}{\overset{|}{CH}}-O- \right]_X \quad \left[-CH_2-CH_2-O- \right]_Y$$

The Polyol

For the most part, the polyol gives the PUR its chemical nature, especially when TDI is the isocyanate inasmuch as the polyol is the major constituent. The secret to making even softer foams is to change the length of the polyol chain.

Two types of polyols are typically used, polyesters and polyethers. The polyesters are usually based on adipic acid, but others are available. The polyethers are derivatives of ethylene and propylene oxides.

The following is a typical polyester (Figure 1.4):

These are essentially hydrophobic chemicals and therefore lead to hydrophobic PURs. The structure of the polyethers is as follows (Figure 1.5):

Polypropylene glycol (left) is essentially hydrophobic, while polyethylene glycol is hydrophilic and is the basis for the hydrophilic PUR business.

The propylene-based polyols (left of Figure 1.5) are currently the basis of most conventional PURs. The methylene group on the polypropylene molecule (at useful MWs) renders it hydrophobic. Contrast this to the polyethylene glycol (right of Figure 1.5), which is water soluble at high MWs. As we said, it is the polyol of choice for most hydrophilic PURs. Both polyols are available in several MWs and the number of –OH groups. This gives the researcher multiple degrees of freedom.

In current practice, foam manufacturers prefer polyethers for the following reasons:

- Lower cost
- Better hydrolytic stability
- Mechanical flexibility

Cross-Linking

Cross-linking is used to control many of the mechanical properties of the final product. Trifunctional alcohols are used for this purpose but any molecule that has more than two reactive sites will do. Cross-linkers for this discussion are typically another polyol. The polyols we have discussed are alternatively called alcohol-capped polyols but in fact they are diols. Cross-linkers in the sense of

Table 1.1 Effect of functionality.

Average functionality	Foam application
2.00	Elastomer
2.07	Carpet-backing foam
2.12	Soft, integral skin foams
2.21	Automobile cushions
2.49	Semirigid foams
2.70	Rigid foams
3.00	Construction grade rigid foams

Source: Wood [1]. Reproduced with permission of John Wiley & Sons.

this argument are small molecules that have three or more alcohol caps. Their effect is to strengthen the molecule by creating more isocyanate bonds.

They have an important physical effect. Without some amount of cross-linking (<5%), foaming will not occur. The cross-linking plays the role of a gelling agent, trapping CO_2 (see section on "The Water Reaction") in the matrix. Without the gelling effect any gases produced would escape leaving a semi-elastomeric product behind.

The average number of –OH groups can be chosen, and this can lead to a certain controlled amount of cross-linking. A component can be added to the prepolymer reaction to develop cross-linking. This has the effect of increasing the number of –OH sites with which the isocyanate can react. This is typically the least expensive way to develop cross-linking.

The primary method of control, however, is the choice of the degree of functionality of the polyol (number of –OH per molecule) whether this is done with a single polyol or by adding another, typically a low MW polyols. Typical additives to induce cross-linking are triols like trimethylol propane (TMP), which is considered a hard segment. Very small amounts are needed for soft foam.

An alternative term used for this effect is the functionality. If the functionality is two (a diol), an elastomer results. If, by the addition of a cross-linker, the functionality is greater than two, foaming occurs (Table 1.1).

The Water Reaction

The last reaction we need to discuss is that of the isocyanates with water. Water reacts with an isocyanate to produce an amine and carbon dioxide gas (Figure 1.6). This is the basis of the PUR foam business. Even with hydrophobics, water is used to create foam, even if much less water is added.

$$\underset{\text{(The isocyanate)}}{R{-}N{=}C{=}O} + \underset{\text{(Water)}}{HOH} \longrightarrow \underset{\text{(An amine)}}{R{-}NH_2} + CO_2\uparrow$$

Figure 1.6 The reaction of isocyanates with water.

$$R{-}N{=}C{=}O + R'{-}NH_2 \longrightarrow R{-}\underset{H}{\underset{|}{N}}{-}\overset{O}{\overset{||}{C}}{-}\underset{H}{\underset{|}{N}}{-}R'$$

Figure 1.7 The reaction of isocyanates with amines (urea linkage).

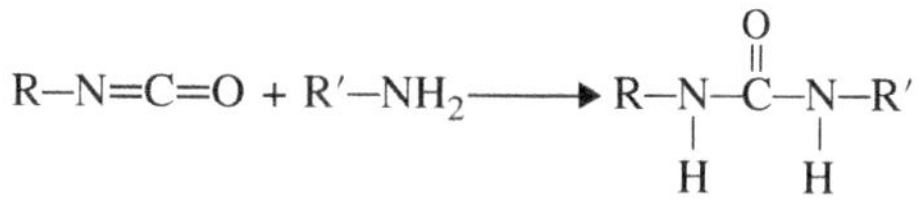

If one wants an elastomer, one must carefully ensure that there is no water in the polyol. To some degree, the amount of water controls the density of the resultant foam. A typical furniture foam might add from 0.5 to 5% water to the formulation. We will discuss this further when we review the processes. Hydrophilic foam formulations can use more water than the polyols. The reaction with water does not end there. You see from the reaction in the previous figure that there is also an amine coproduct. The amine reacts with an isocyanate to produce a urea linkage (Figure 1.7).

It is the water reaction and the amine reaction that results in PUR foam. A foam manufacturer needs to be aware of both reactions. While the production of CO_2 is the driving force, unless the amine reaction proceeds, all the CO_2 would be lost to the atmosphere. It is the amine reaction coupled with a polyol with a functionality greater than 2 that causes the reacting mass to gel up. In the industry this is called cream time, but for the chemist, the effect is the generation of a three-dimensional matrix that first traps the CO_2. As the mass expands, the internal pressures begin to burst the windows between the cells, thus creating an open-cell foam.

All this happens under close temperature control. We will cover this further when we turn our attention to the special case of hydrophilic PURs. The two reactions (CO_2 and amine) have different activation energies. Higher temperatures favor the CO_2 reaction that, if high enough, causes the foam to collapse. The internal pressure is high enough to overcome the strength of the gel and CO_2 escapes.

Thus we have described the simultaneous reactions of polymerization and expansion. The juxtaposition of these two reactions is the basis of the PUR foam industry. However, it also has other implications. Consider the commercial of an adhesive brand, an MDI-based prepolymer with a proprietary polyol. We can guess that it is highly cross-linked. We can also guess that it uses a hydrophobic polyol. Herein lies a paradox. In advertisements it claims and by our experiments confirms that it is "waterproof" yet it is activated by water. Another property is that it expands to fill, for instance, cracks. A dramatic video in their website shows the bonding of a wood gate to a ceramic post. The animation shows the curing adhesive penetrating both the wood and the ceramic. The question one must ask as a concerned consumer is how all this can be true. The answers are in the discussion of the chemistry given previously.

We know that a prepolymer can be formulated with excess isocyanate and a highly cross-linked hydrophobic polyol. This is then activated by adding a small amount of water. If not confined in a mold, the activated prepolymer will foam, probably developing closed cells. When the reaction finishes, what remains is a hydrophobic foam. If the activated prepolymer is in a confined space, the expansion penetrates the pores of the two materials. Once fully cured, the two materials are bound together by a strong but brittle elastomer.

This will be discussed further in the process sections but for now it serves as an introduction to process control.

Process

There are two dominant processes by which the chemistries discussed previously are converted into useful products. The first of those combines all the raw materials in the formulation and allows them to react to form the elastomer or foam. In the second process the polyol is capped with the isocyanate and isolated as an isocyanate solution in the polyol. This is the process used for all hydrophilic PURs. We will discuss both processes starting with the more important of the two, the so-called one-shot process.

The One-Shot Process

As chemists we are accustomed to large stirred tanks to which a sequence of chemicals is added and a complex program of heating and cooling. Conventional PUR foam is literally made by slamming together all of the components discussed previously plus some others and then deposited on a conveyor. I use the word slam not without justification. The technical term is impingement, but the effect is the same. In the late 1950s, it was discovered that one could manufacture PUR directly from the component parts using certain surfactants and catalysts. In the one-shot technique, as it is called, a polyol blend is made containing the surfactant, the catalysts, the blowing agents, and the other components. This blend is then quickly and intimately mixed with an isocyanate phase in what is called an impingement mixer. The fluids are forced together through nozzles, thus creating an emulsion. The emulsion is placed in a mold or other receptacle where the foaming reaction proceeds. Figure 1.8 shows the essential parts of an impingement mixer. As the plunger is withdrawn, the two streams are "slammed" together. Note the several process streams entering the mix head.

The following is a typical formulation used in the one-shot process (Table 1.2):

Depending on the amount of water, this formulation list can be used to make elastomers or low-density foams. In the case of elastomers, the emulsion

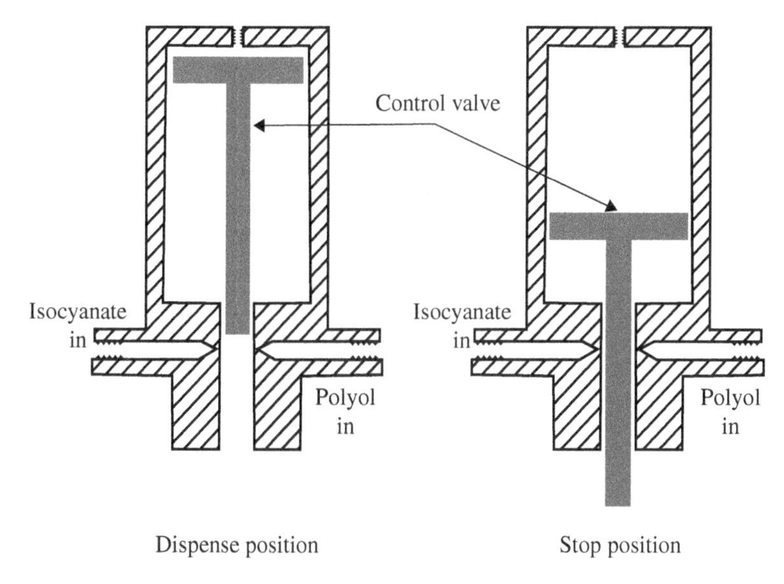

Figure 1.8 An impingement mixer.

Table 1.2 Component list for a conventional polyurethane by the one-shot process.

Component	Parts (mass)
Polyol	100
Water	1.5–7.5
Fillers	Variable
Silicone surfactant	0.5–2.5
Amine catalyst	0.1–1.0
Tin catalyst	0.5
Chain extenders	0–10
Cross-linker	0–5
Isocyanate	25–85

Source: Adapted from Herrington and Hock [2].

might be injected into a mold or caste into a sheet. For foams, the emulsion is deposited on a moving conveyor lined with a paper release liner. At this point a time line is set. After a short delay the water reaction begins liberating carbon dioxide and producing the amine coproduct. Shortly thereafter the polyol/isocyanate/amine reaction begins, which increases viscosity first but

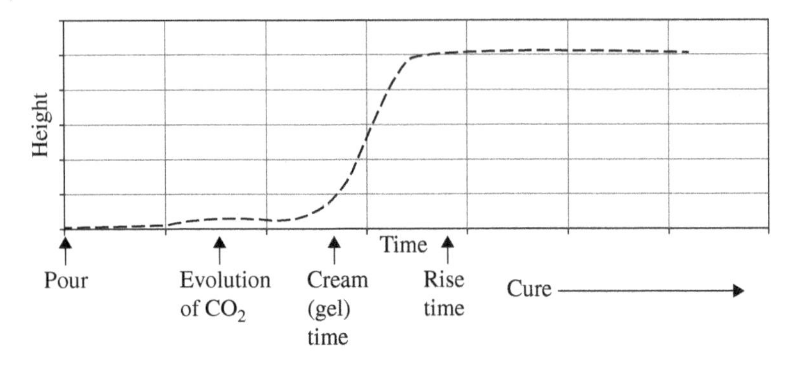

Figure 1.9 The foaming process after leaving the mix head.

then begins to develop a gel structure. In the industry this is called cream time; in fact it is a cross-linked molecular structure capable of trapping the carbon dioxide. The gel matrix continues to increase in strength as more and more carbon dioxide is produced. It is important to note that each of these reactions has different activation energies. Thus temperature affects the reactions at different degrees. In a controlled process, without actually knowing what is going on, the operator is aware of the temperature of the components and the exotherm produced during the reaction. This determines the density and structure of the resultant foam. If properly controlled, as the foam is rising, complex but predictable cell structures separated by windows develop. As the density decreases, those windows become thinner. This happens while the internal pressure created by the carbon dioxide gets high enough to break the windows.

The process is the same whether we are discussing the one-shot process or the prepolymer process to follow and so it is appropriate to use the following graphic that summarizes the foaming process (Figure 1.9).

The result is a stable foam structure. The properties, as we have discussed, depend on the components and temperature of the emulsion. The density of a commercial foam is around $2\,lb/ft^3$. If the thermal conditions are correct, the result is an open-cell foam, the most common PUR foam.

The Prepolymer Process

The first PURs were made by what has come to be known as the prepolymer process. In the 1950s catalysts and surfactants were developed that made the one-shot process the preferred technique for foam production.

The prepolymer process involves the manufacture of an "isolated intermediate" that can be stored and sold as a product. Upon exposing the product to a polyol or, in the case of hydrophilic foam, water, the reaction proceeds to make a foam or an elastomeric coating. Prepolymers have experienced a resurgence with the development of PUR paints, coatings, and adhesives. The big advantage of these formulations is they can be produced to have low viscosity and cure in atmospheric moisture. In the case of paints and coating, this technology eliminates the need for volatile organic solvents.

In many prepolymer processes, a small stoichiometric excess of diisocyanate is reacted with the polyol. Again, if the product is to make foam, the polyol has a net functionality greater than two. In the case of hydrophilic PUR, this is accomplished by using polyethylene glycol and a few weight percent trimethylol propane. Water is carefully removed from the reaction mixture, thus ensuring that the reaction occurs between the isocyanate and polyol only. The reaction is conducted between 60 and 120°C. As the isocyanate and polyol react, the viscosity begins to increase. If the prepolymer is to be used for a coating, lower viscosities are desired. For adhesives, a higher viscosity is preferred. For hydrophilic foams, it has become typical that the viscosity of the prepolymer be from 10 000 to 18 000 cps. In any case, at a specified viscosity, the reaction is stopped by cooling. The product is a diisocyanate-rich solution of a polyol capped with an isocyanate. You will note that the prepolymer is composed of only urethane linkages. When a prepolymer is exposed to water, carbon dioxide is released and amine end groups are formed, restarting the reaction and bringing it to completion. The water can be as little as atmospheric moisture (for an elastomeric coating or adhesive) or in the case of hydrophilic PURs can be as much as one to three times the mass of the prepolymer (Figure 1.10).

All hydrophilic PURs are made from prepolymers. Several companies sell them and while there is little variability in their specifications, one can choose the level of cross-linking. There are TDI versions and MDI products. Each has unique characteristics that need to be considered.

As with the one-shot process, there are specially designed pieces of equipment made for producing foam. In the one-shot process, the equipment is referred to as "high pressure meter mix" because of the force needed to "impinge" the ingredients. In the prepolymer process to make foam, "low pressure" equipment is used.

Specific to commercial hydrophilic prepolymers, one to three parts of water with a surfactant is mixed in a low pressure mixer and then deposited into a mold or onto a conveyor.

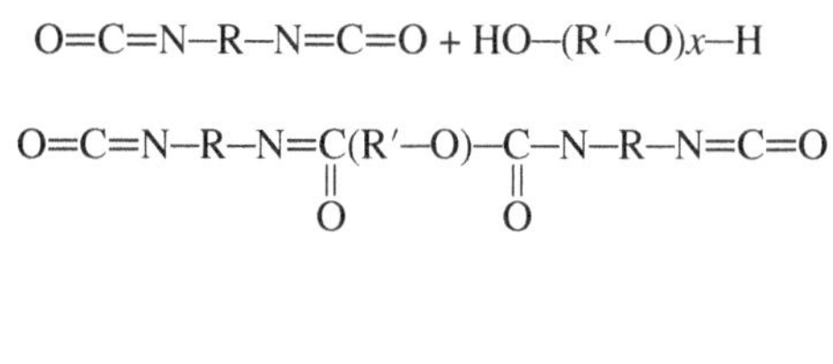

Figure 1.10 The prepolymer reaction (the triol is not shown for clarity).

O=C=N–R–N=C=O + HO–(R′–O)*x*–H

$$O{=}C{=}N{-}R{-}N{=}\underset{\underset{O}{\|}}{C}(R'{-}O){-}\underset{\underset{O}{\|}}{C}{-}N{-}R{-}N{=}C{=}O$$

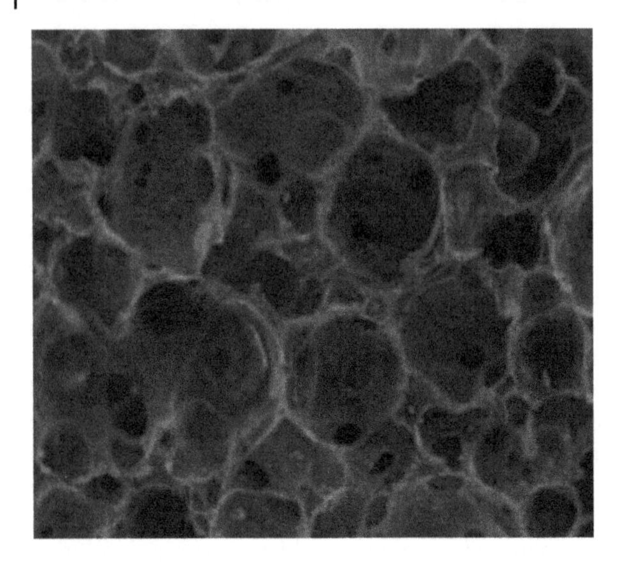

Figure 1.11 Micrograph of an open-cell foam.

We will explore this to some degree in our research into polyethylene/polypropylene block copolymers. This research was not focused on commercial applications but rather an effort to expand the possibilities. Periodically we will refer to this as a hydrogel. We use an unofficial definition of a hydrogel as a polymer that is capable of absorbing 20+% of its weight in water. Note that this assigns human flesh and all internal organs as being hydrogels or composites thereof.

Regardless of whether one uses the one-shot process or a prepolymer, if a foam is to be the product, the reaction goes through the stages shown previously, that is, emulsion, gelation, foaming, and curing. The product is typically an open-cell foam (Figure 1.11).

The degree of openness and whether it is open at all is controlled in the process, but most PUR foams are open to some degree. We will discuss these in the chapter on laboratory practices.

Post Processing

For the purposes of this discussion, most conventional PUR foams result in an open-cell structure. We refer to this as furniture foam. While chemically capable of serving as an immobilizing material (by adsorption of cells, for instance), it is architecturally inappropriate. One could not reasonably expect to pass fluids through it, including air. Reticulation is a post-processing step that is used to improve this, however, and is the subject of this section. It is used when the foam is to be used for air filters and products that depend on the free flow of fluids through it. It is clear upon examining the micrograph of the open-cell foam that while open to have air pass through it, the pass is "tortuous" and this leads to a significant pressure drop. This will become clear when we describe the reticulation process. The intent is to further remove the "windows" that separate the cells.

There are two processes to accomplish this. The first process was to immerse the foam in a hot sodium hydroxide solution. It is effective and produces a unique surface chemistries. This process is being phased out in favor of the now dominant process of zapping the foam. In this process, the foam is place in a chamber. Hydrogen and oxygen are then piped in and a spark ignites the gasses, thus burning the windows between the cells.

Both processes create a foam structure of remarkably lower pressure drop as you might guess by the micrograph that follows (Figure 1.12).

It is important to remember that this foam was made from open-cell foam. In manufacturing facilities, however, the open-cell foam is very carefully made to develop an open-cell structure that is uniform, as opposed to furniture foam in which the cells have a broad distribution of pore sizes.

The material is commercially available in a wide range of pore sizes. As the pore size decreases, the surface area and pressure drop increase.

We will spend some time on describing this specialty PUR foam because it plays an important role in the current and future applications of hydrophilic foam. Conventional hydrophilic foams, as we will describe them, are open-celled. As such they are typically high in pressure drop. Even when formulations are used to make a more open structure, when wet they are not strong enough to allow high flow rates through them. Using a reticulated foam as a substratum and grafting a hydrophilic PUR to the inside surfaces produces a hydrophilic surface on a reticulate structure. Reticulated foams are typically

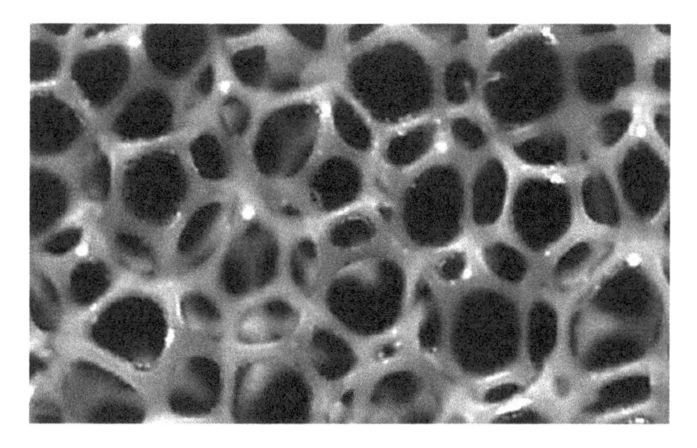

Figure 1.12 Typical reticulated foam.

Table 1.3 Typical properties of a commercial reticulated.

Nominal pore size (pores per inch)	Minimum pore size (ppi)	Maximum pore size (ppi)	Void volume (%)	Internal surface area (m^2/m^3)
100	80	110	98	6900
80	70	90	97	5900
55	55	65	97	3900
45	40	50	97	2800
30	25	35	97	1650
20	15	25	97	980
10	8	15	97	490
3	3	5	97	330

Source: Adapted from FXI Corp. Technical Product Function Sheet, FS-998-F-5M.

designated as to their pores per inch (ppi). Table 1.3 is taken from product literature of FXI, a major producer of this foam.

Notice the uniformity of the cell sizes. This is an important property of reticulated foams as it reflects on the lot-to-lot uniformity and the flow characteristics.

Architecture of Polyurethane Foam

Whether foam is made by the one-shot process or from a prepolymer, once the materials leave the mix head, the process can be thought of as chaotic. You have done what you could to control the physical and environmental aspects. You have adjusted the ratios, components, and temperature. You have provided a vessel into which the liquid is poured and enough physical energy in the form of mixing to create an emulsion of a consistent texture. Once the emulsion leaves the mix head, however, there is little or nothing you can do. While you have a few seconds before cream time to add more liquid, the polymerization and CO_2 generation must be done without being disturbed. The mass will expand and develop a structure that is not influenced by external effects. One can add nucleating components or emulsified air that initiates a CO_2 bubble, but even then, it is as if the PUR architecture is developed by an “invisible hand.” Microscopic examination of free rise foam shows that there is a consistency. This is most easily seen with reticulated foams, but it is true of furniture grade as well. The rising foam by thermodynamic processes assumes the lowest energy state. Each cell in the foam must adjust itself to the cells that surround it. The bars and struts, of which the foam is constructed, must have a uniformity of stresses in order to develop the bulk properties (tensile strength, compressibility, etc.).

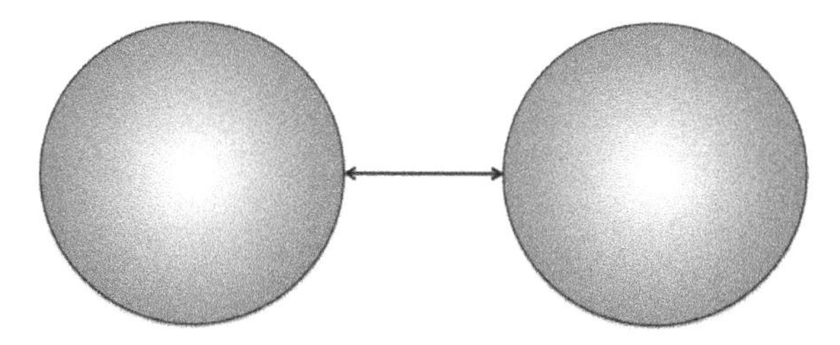

Figure 1.13 Two spheres approaching one another.

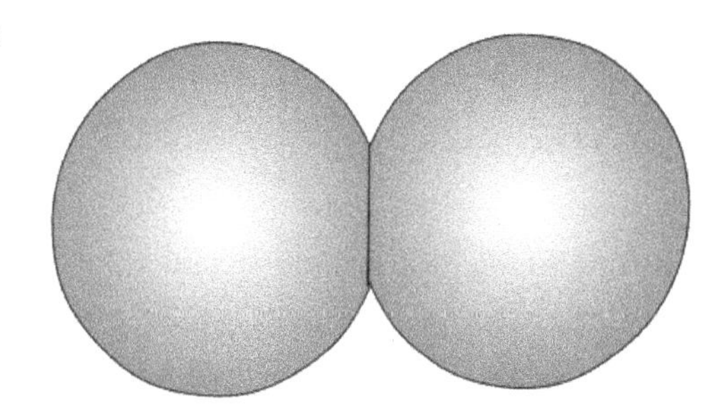

Figure 1.14 Lowest surface energy among two cells.

It will not be surprising that this "natural structure" has been the subject of mathematical investigation, not directly on PUR foam as the treatment was developed before the invention of PUR. The target of the investigation was to determine an ordered structure for bubbles of the same cell size (monodispersed). We will discuss the uniformity of pore sizes in a PUR foam and I think you will be surprised.

To explain further, consider two spheres approaching another. The shape of each is governed by Plateau's rule of minimum surface energy (Figure 1.13).

As the spheres come into contact, the surface energies must adjust to each other. The lowest energy is to form a window between the cells (Figure 1.14).

Now imagine the same process occurring simultaneously in three dimensions. All interactions must yield a flat surface between the cells. If cells were cubes, the mathematics would be simple. Combining spheres are mathematically complex.

It is the invisible hand as described by the Plateau rule and the minimalization of surface energies. Relating it back to PUR, it is logical to assume that the chaos describing the building of a PUR foam would follow this invisible hand and indeed it does. Again this is best seen in reticulated foams. We have used this micrograph before but it is worth repeating in the context of this discussion.

In 1887, Thomson proposed that the tetrakaidecahedron was the best approximation [3]. It was not perfect but it was the best fit, that is, the lowest energy. As it happens, constructing a single cell reveals a bowing of some of the components. The structure is built from six squares and eight hexagons. The following micrograph "confirms" the theory (Figure 1.15).

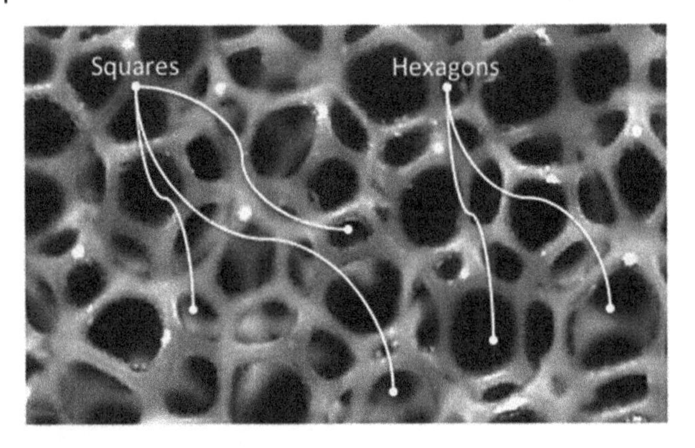

Figure 1.15 Foam architecture.

Grafting to the Polyurethane Foam

We must change our perspective from a PUR structure to PUR as a surface treatment. We are thinking of hydrophilics, specifically inasmuch as in systems we discuss water is the fluid. While this is an important factor, making chemical adjustments to the surface by immobilization are usually done on hydrophilic surfaces. Finally, while it is the dominant material in many product markets, it has certain problems that have limited its growth in some important applications. Whether it is in agriculture, cosmetic applicators, or chronic wound dressing, it has become the material of choice due to its ability to absorb and retain moisture. The typical hydrophilic PUR will absorb as much as three times its weight in water. While not a so-called superabsorbent, its physical properties give it great value. In a sense, it is a convenient way to immobilize water and solutes therein.

From a physical point of view, it has some deficiencies. Applications that require anything more than minimal stress (tensile or compression) are beyond the materials capability, especially when wet. Moreover, we went through a discussion of improving the architecture of open-cell foam to open up its structure (reticulation). This technique is not available to hydrophilic foams. While there are techniques to open the cell structure to some degree by the use of surfactants, the strength of the foam counteracts these techniques with regard to mass transport.

Hydrophilic PUR foams are not available in low density. Commercial hydrophilic foams are in the range of 6 lbs/ft^3 compared with 2 lbs/ft^3 for conventional PU foams. Higher density results in higher cost and this make hydrophilics among the most expensive in the category.

In order to mitigate the problems with strength and cell structure, we investigated the concept of coating a more structurally appropriate substratum.

By grafting hydrophilic PUR foam onto another surface, these problems are all but eliminated. In our work, reticulated foam was the material that interested us the most. You should be aware that several patents were developed by the author that describes these composites [4]. Having said that, our purpose in discussing it here is to illustrate the use of hydrophilic PUR, specifically as a surface coating. In later chapters we will add a discussion of how, in one step, biomolecules can be covalently bonded to this surface. Thus, if successful, a material could be developed with optimum architectural properties, a hydrophilic surface for biocompatibility, and finally a convenient surface for immobilization of bioactive molecules.

Hydrophilic PUR starts with polyethylene glycol as opposed to the polypropylene glycol commonly used in conventional PURs. Hydrophilic prepolymers are available from a number of companies in the United States and Europe. They are always supplied as a prepolymer in viscosities around 10000 cps. They are clear to amber in color.

The process to make hydrophilic foam is simple, and this is discussed in the chapter on laboratory practices. The prepolymer is emulsified with water in a high intensity mixer. It is then deposited onto a conveyor lined with silicone-coated paper or into a mold. After that, the process follows the steps common to all PUR foam.

The process to "graft" hydrophilic PUR to a reticulated foam is described here. Once the water/prepolymer emulsion is made, it is immediately deposited onto a moving sheet of reticulated foam. The equipment to accomplish this is shown in Figure 1.16. Very low coating weight (5–10%) is produced by a different process described in the patents.

After about 10 min, the material is substantially cured and can be rolled. The amount of coating is determined as the percent hydrophilic PUR on

Figure 1.16 Coating process for the hydrophilic polyurethane composite. (*See insert for color representation of the figure.*)

the composite. We are not sure there is a covalent bond between the hydrophilic PUR and the hydrophobic PUR substratum or scaffold. What we do know is that it is durable. We have never experienced a delamination. Coating rates can be as low as 5%. This is apparently sufficient to render the surface of the composite hydrophilic. Figure 1.17 shows a composite with a coating weight of 25%.

While our research is on reticulated foam as a substratum, the point of this discussion is to illustrate the use of PUR as a coating as opposed to a stand-alone structure. It would appear to be a convenient way to develop a hydrophilic surface to another material. Our purpose here is to describe hydrophilic PUR as a surface treatment. Of course not all materials will take a hydrophilic coating. In those cases an intermediate layer or surface treatment is needed. In the chapters that follow, we will describe other materials as options as a substratum for immobilization. We will make the case that hydrophilic PUR coatings offer an appropriate and convenient surface for immobilization as well as an effective architecture.

While the remainder of this discussion deals more broadly with PUR hydrophilics, the composite discussed previously offers advantages. By way of example, enzymes are conveniently immobilized using a prepolymer, but the temperature must be low enough to prevent denaturation. The result however is a foam with poor mass transport properties. That is, the accessibility of the

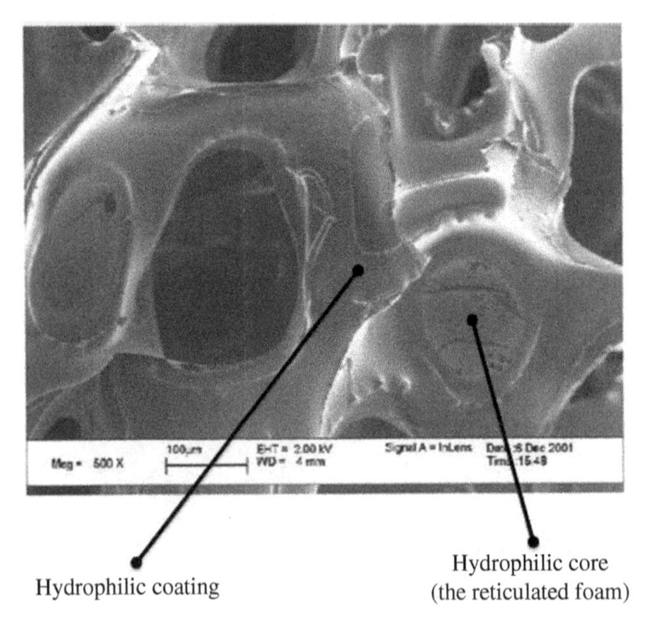

Figure 1.17 Micrograph of the composite with a coating weight of 25% hydrophilic PUR.

enzyme is less than ideal. By doing the reaction on the surface of a material with good transport properties, the accessibility is improved.

Still further, while hydrophilicity is considered inherently biocompatible, consider cell adhesion as an aspect to be considered. While not PUR, it does speak to hydrophilicity. Poly(*N*-isopropylacrylamide) (PIPAAm) was grafted onto the surfaces of commercial polystyrene cell culture dishes [5]. The surface was shown to be hydrophobic at temperatures over 32°C and hydrophilic surface properties below that temperature. Endothelial cells and hepatocytes attached and proliferated on PIPAA-grafted surfaces at 37°C. The optimum temperature for cell detachment was determined to be 10°C for hepatocytes and 20°C for endothelial cells. Cells detached from hydrophobic–hydrophilic PIPAAm surfaces not only by reduced cell surface interactions but also by morphological changes.

Researchers at Medtronics (Santa Rosa, CA, USA) studied polymer coatings used for the delivery of drugs from a vascular stent [6]. They determined that it is critical to balance the hydrophilic and hydrophobic components of the polymer system to preserve biocompatibility and to maintain controlled drug elution. In their study, hydrophilicities of the polymer surfaces were determined by contact angle measurements. Biocompatibility was evaluated by an *in vitro* assay system in which activated monocyte cells were exposed to the polymer. Polymers of a more hydrophobic nature were observed with enhanced monocyte adhesion, whereas those that were more hydrophilic did not induce monocyte adhesion. The report supports the hypothesis that polymer composition is a feature that determines *in vitro* biocompatibility. The results of the study suggest hydrophobic, but not hydrophilic, polymer surfaces support adhesion of activated monocytes to the polymer scaffold.

Biodegradable PUR

While we see an important role for extracorporeal bio-scaffolds, much of the research is on implantable and degradable polymers. A critical factor in this is that a scaffold can be a temporary device to guide the spreading of cells, *in vivo*, until the development of an extracellular matrix and possibly vascularization. At that point the scaffold should degrade to resorbable or at least nontoxic residues. Timing is everything in this process and that presents a significant problem. Degradation to neutral fragments is a study in itself.

In this text our primary focus is on the scaffold, without regard to its fate *in vivo*. We do not ignore the fact but given our focus, what we discuss is more appropriate to an extracorporeal device. In later sections we will discuss a "liver model" as a capsule outside the body as the first step (and maybe the final step) in developing a liver support device. We will discuss at length a research done, in our opinion, on the scaffold as opposed to chemistry in Japan.

They chose a material based on architecture to the exclusion of compatibility issues. They conducted small animal studies with, again in our opinion, with great success. It was necessary, however, to pass plasma through a reticulated foam colonized with hepatic cells. The foams were hydrophobic, and therefore blood contact was not possible.

Nevertheless, development of a biodegradable scaffold as described previously is a logical target for investigation. Where we differ in our approach is that an implantable device combines the structure and the chemistry. In our work we focus on the scaffolds alone, leaving the chemistry of degradation to subsequent studies. We wanted to know what the scaffolds look like and then give them appropriate chemistry.

It would not be appropriate to ignore the fate of a synthetic material, like a scaffold, upon implantation. Therefore to complete the discussion of chemistry, we will talk about how one builds degradability into the molecule.

For review, hydrophilic and other PURs are made up of a number of linkages. For conventional polymers, the aromatic ring(s) that combines the parts are not considered degradable in the normal sense. Urethane linkages are formed by the reaction of isocyanates with hydroxyl-functional molecules, while urea linkages are formed by reaction with amines. It is these bonds that begin the process of *in vivo* degradability. Alternative isocyanates, the hydroxyl-functional molecules and the reactions with amines are all subject to examination for potential degradation sites. The goal is to unzip the polymer into fragments.

The degradation of PURs has been the subject of its use in agriculture. One of the early applications of hydrophilic PURs was as a growth medium for plants. While it has other functions, its primary role was as a binder for non-soil materials like peat moss and bark ash. It has become an important technology, especially in high value plants. We assisted in the spread of the technology in Europe and the United States. Our lab supervised the planting of trees and ornamentals in Sweden, Canada, and, of course, here. International Horticultural Technologies in Hollister, CA, is a leader in this business.

Early in the project, the issue of biocompatibility came up. People were concerned about the release on toxic substances into the soil. As there was no evidence of toxic effects, a policy we promoted was the idea that the PUR broke apart (unzipped) and after a short period became "indistinguishable from soil."

Nevertheless, we approached the University of Illinois to conduct a full degradation study. The goal was to see how far the degradations would go. Complete degradation would be the conversion of the carbon and nitrogen to biomass with water as a coproduct. I left before the study began, but in the investigation, the fate of hydrophilic PUR was to be identified by gases released and biomass developed. I don't believe that the study was ever completed, but that type of study at some point would have to be done on *in vivo* degradation.

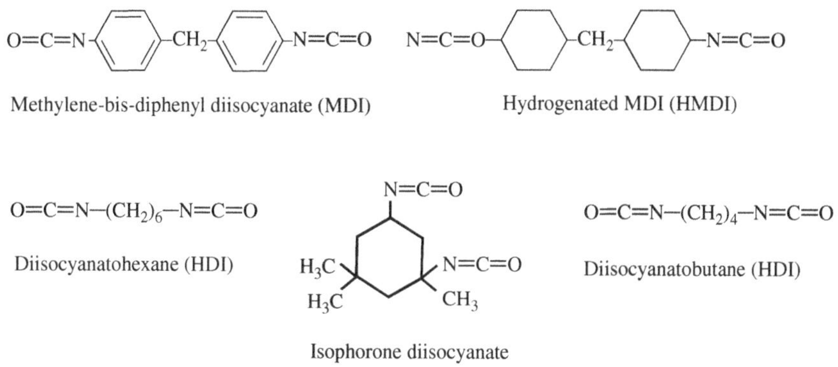

$O{=}C{=}N{-}C_6H_4{-}CH_2{-}C_6H_4{-}N{=}C{=}O$

Methylene-bis-diphenyl diisocyanate (MDI)

$N{=}C{=}O{-}C_6H_{10}{-}CH_2{-}C_6H_{10}{-}N{=}C{=}O$

Hydrogenated MDI (HMDI)

$O{=}C{=}N{-}(CH_2)_6{-}N{=}C{=}O$

Diisocyanatohexane (HDI)

N=C=O, H_3C, H_3C, N=C=O, CH_3

Isophorone diisocyanate

$O{=}C{=}N{-}(CH_2)_4{-}N{=}C{=}O$

Diisocyanatobutane (HDI)

Figure 1.18 Isocyanates appropriate for biodegradable polyurethanes.

Returning to a more general discussion of degradability, a review of the parts of PUR with regard to degradability is appropriate. As we said, the aromatics are not degradable, but aliphatic isocyanates can be. Figure 1.18 shows a few of the options available to the chemist for consideration. Hydrogenated MDI (HMDI), hexamethylene diisocyanate (HDI), and 1,4-diisocyanatobutane (BDI) are aliphatic. Lastly, isopherone diisocyanate is also aliphatic. Polyurethanes prepared from aliphatic isocyanates have been reported to biodegrade *in vitro* and *in vivo* to non-cytotoxic decomposition products [7]. Foams prepared from the aromatic 2,4-TDI (not shown) were reported to degrade under simulated physiological conditions to the toxic 2,4-toluene diamine. In the 1970s and 1980s, the failure of PUR-covered silicone breast implants generated concerns about the safety of biomedical devices incorporating PURs [8]. One study reported that the degradation products from the PUR foam included acutely toxic, carcinogenic, and mutagenic aromatic diamines [9]. However, whether the concentrations of these toxic degradation products can reach physiologically significant levels *in vivo* was inconclusive and has not been resolved. As an aside, the structure of the implant was a "balloon" filled with a silicone fluid in a silicone package. Because in-growth by tissue was necessary to stabilize the device, a hydrophilic PUR foam (not Hypol, however) was applied to the outside of the breast implant. We were asked to quote on supplying a similar product to a European manufacturer but declined as this was a major concern. In the same report [2], biodegradable PURs synthesized from the aromatic MDI, however, have been shown to biodegrade *in vivo* to non-cytotoxic decomposition products.

Turning to the polyol section of the polymer, examples of likely polyols used in the synthesis of biodegradable tissue scaffolds are shown in Figure 1.19.

With regard to the polyols, the ethylene and propylene glycols (PEG and PPG, respectively) are the basis of most commercial PUR. While we expect

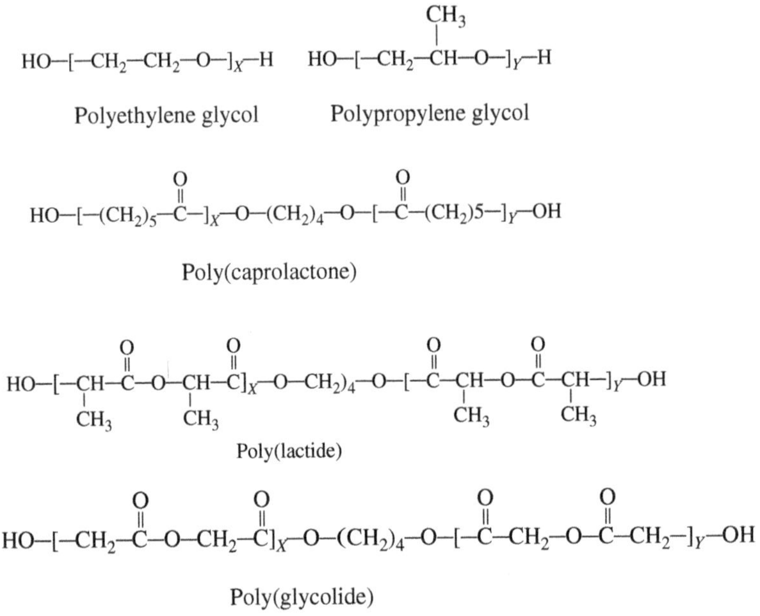

$$\mathrm{HO{-}[{-}CH_2{-}CH_2{-}O{-}]_X{-}H} \qquad \mathrm{HO{-}[{-}CH_2{-}CH(CH_3){-}O{-}]_Y{-}H}$$

Polyethylene glycol Polypropylene glycol

$$\mathrm{HO{-}[{-}(CH_2)_5{-}C({=}O){-}]_X{-}O{-}(CH_2)_4{-}O{-}[{-}C({=}O){-}(CH_2)5{-}]_Y{-}OH}$$

Poly(caprolactone)

$$\mathrm{HO{-}[{-}CH(CH_3){-}C({=}O){-}O{-}CH(CH_3){-}C({=}O)]_X{-}O{-}CH_2)_4{-}O{-}[{-}C({=}O){-}CH(CH_3){-}O{-}C({=}O){-}CH(CH_3){-}]_Y{-}OH}$$

Poly(lactide)

$$\mathrm{HO{-}[{-}CH_2{-}C({=}O){-}O{-}CH_2{-}C({=}O)]_X{-}O{-}(CH_2)_4{-}O{-}[{-}C({=}O){-}CH_2{-}O{-}C({=}O){-}CH_2{-}]_Y{-}OH}$$

Poly(glycolide)

Figure 1.19 Polyols appropriate for biodegradable polyurethanes.

that they will play an important role in tissue engineering in the form of a scaffold template, for the purpose of this discussion, we have to consider them biodurable. This is true for the polyester versions. Being biodurable does not say that they are inert, *in vivo*. Monocytes are recruited to the surface of implant, where they can differentiate and develop foreign body giant cells [10]. The release of biologically active molecules causes surface defects and loss of mechanical strength. They are not degraded into smaller molecules, the goal of what we refer to as biodegradation.

Inasmuch as they form the backbone of the PUR industry, they must fill multiple physical and mechanical roles. Thus both PEG and PPG are available in a broad spectrum of MWs and degrees of functionality. In addition there are families of surfactants based on block polymerizations of PPG with end caps of PEG. Among those products is the Pluronic family of products (BASF Corp.). They are available in various MWs and mass percentages of PEG. They can be used to make PURs, some with useful properties. We will illustrate with a study of a replacement for degraded spinal disc.

For degradable polyols, higher MWs are needed for mechanical strength and elasticity. Lactic acid and glycolide polymers, while good candidates because of their biodegradability, are of low MWs. Copolymers of polyethylene oxide and polylactic acid, for instance, require what are called chain extenders to increase their MW. Many researchers have taken this route.

It is critically important to realize that the polyols in the table are probably sufficient for most research; the researcher should not be limited to this list. While there is significant research that needs to be done with those polymers, chemistries outside have shown interesting and probably important research.

By way of example, researchers at the University of Alabama worked to develop a peptide-modified PUR to enhance endothelialization for small diameter vascular graft applications [11]. YIGSR peptides were incorporated into the polymer backbone. Endothelial cell adhesion, spreading, proliferation, migration, and extracellular matrix production were improved compared to controls. Additionally, competitive inhibition of endothelial cell attachment and spreading was found when cells were incubated in the presence of soluble YIGSR peptides. The incorporation of the peptides into the polymer backbone did not significantly affect the tensile strength. However, the elastic modulus was decreased, whereas elongation was increased.

The point is that while the choices of isocyanates are limited, the choice of soft segments is not. Functionally this is the portion of the molecule that makes the scaffold possible.

Mechanism of Biodegradation

For review, urethane prepolymers are formed by the reaction of isocyanates with hydroxyl-functional molecules. This creates a network of urethane linkages. Hydrophobic PUR foams are made by adding additional polyols, catalysts, surfactants, and fillers and a little water. The reaction creates more urethane bonds. Water reacts with isocyanates to form amines and carbon dioxide gas. The reaction of the amine with an isocyanate results in urea bonds. Both reactions, however, result in an increase of MW. Thus the reactions result in a polymer composed of urethane, urea, and, depending on the polyol, a backbone that is either biodurable or biodegradable. In the later case, let us examine how those bonds are affected *in vivo*.

Biodegradable PURs are designed to undergo hydrolytic or enzymatic degradation to non-cytotoxic decomposition products *in vivo* (Figure 1.20).

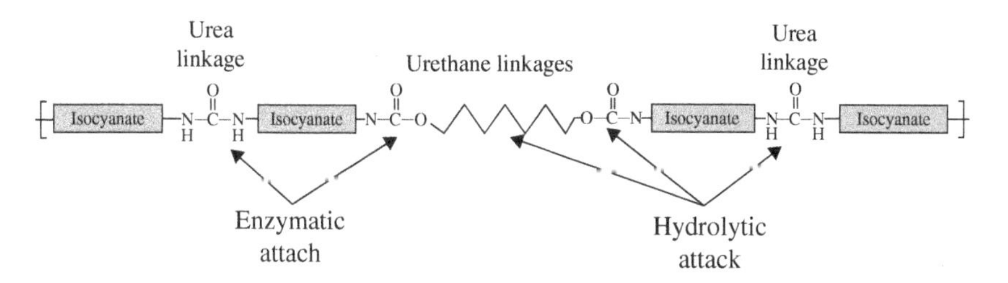

Figure 1.20 *In vivo* degradation of polyurethanes.

Mechanisms of hydrolytic degradation have been suggested in the literature [12]. Ester linkages hydrolyze both *in vivo* yielding urethane and urea and other fragments. The composition of the polyester polyol component of the polymer has been shown to control the degradation rate *in vitro*. Polyurethanes with hydrophilic soft segments suggest an increase in the degradation rate. Depending on the isocyanate used to synthesize the PUR, additional degradation of urethane and urea fragments to free polyamines has been reported, but there is a lack of consensus in the literature on the extent of hydrolysis of urethane and urea groups. Hydrolysis of the ester group in lysine polyisocyanates yields a carboxylic acid group. It has been reported that urethane and urea linkages are only enzymatically degraded [13].

It is generally, yet not exclusively, thought that lysine-derived polyisocyanates biodegrade *in vitro* and *in vivo* to non-cytotoxic decomposition products.

More Examples

To illustrate the technology of biodegradable PURs, we have selected a number of examples of synthesis followed by in the minimum *in vitro* studies. The first used the poly(ϵ-caprolactone) and polyethylene/polypropylene glycol copolymers.

Linear biodegradable PURs were synthesized based on poly(ethylene oxide-propylene oxide-ethylene oxide) block copolymer and poly(ϵ-caprolactone) [14]. The polymers absorbed up to 3.9% of water depending on the chemical composition. The tensile strength and elongation at break were in the range of 11–46 MPa and 370–960%, respectively. The glass transition and soft segment melting temperatures were measured. Degradation *in vitro* caused 2% mass loss and 15–80% reduction of MW at 48 weeks. The extent of degradation was dependent in part on the hydrophilic content. The materials containing the block copolymer degraded more. Degradation caused insignificant changes to the pH of the medium.

Biodegradable synthetic polymers for tissue-engineered products and therapies were reviewed by Gunatillake *et al.* [15]. Synthetic polymers were discussed with regard to synthesis, properties, and biodegradability. Degradation modes and products were summarized. Polyesters and their copolymers, PURs, and acrylate/urethane systems were discussed. Polyesters such as polyglycolides, polylactides, and their copolymers still have a prominent position in the field, although the release of acidic degradation products, processing difficulties, and limited range of mechanical properties remain as major disadvantages. Injectable polymers based on urethane and urethane/acrylate are shown to have promise as systems for tissue-engineered products.

In a study that involved a formulation, PUR may have significance to our theme and so it is included here as a curiosity. Fatty acid urethane derivative of dehydroepiandrosterone (DHEA) was evaluated as an additive to increase biostability [16]. The effect of the modified DHEA additive on the biostability of a poly(ether urethane urea) was examined after 5 weeks of subcutaneous implantation in rats. There was no evidence of degradation of the PUR underneath the modified DHEA surface layer as compared with the PUR control. It was assumed that the modified DHEA self-assembled into a protective surface coating that inhibited degradation of the PUR.

Returning to the synthesis for tissue engineering, a cross-linked 3D biodegradable PUR scaffold with controlled hydrophilicity for bone graft substitutes was synthesized from biocompatible reactants. Several scaffolds were made with varying hydrophilic-to-hydrophobic ratios. The main components were hexamethylene diisocyanate, poly(ethylene oxide), poly(ϵ-caprolactone), and water. Calcium carbonate and hydroxyapatite were used as inorganic fillers. The scaffolds had an open-pore structure, the sizes of which depended on the formulation. The compressive strengths were measured and found to increase with increased polycaprolactone and inorganic fillers. The scaffolds underwent controlled degradation *in vitro*. The degradation increased with the increased ratios of polyethylene oxide.

In an interesting study that could have broad implications to the mode of degradation, isocyanates were synthesized from amino acids [17]. The reactions included the preparation of α-isocyanatoacyl chlorides from the corresponding amino acids. The amino acids glycine, L- and DL-α-alanine, L-leucine, and L-phenylalanine were investigated. The mechanical properties of PURs made from the "amino-isocyanates" were measured as well as the degradation behavior.

Not a scaffold, but Jeong reported the synthesis of a biodegradable hydrogel made up of poly(ethylene oxide) and poly(L-lactic acid) [18]. Aqueous solutions of these copolymers exhibit temperature-dependent reversible gel–sol transitions. The hydrogel can be loaded with bioactive molecules in an aqueous phase at an elevated temperature (around 45°C), where they form a sol. In this form, the polymer is injectable. On subcutaneous injection and subsequent rapid cooling to body temperature, the loaded copolymer forms a gel that can act as a delivery system for drugs. While not suitable as a scaffold, it would make an interesting coating.

Conclusion

This is an introduction to the chemistries we will be discussing. We will be describing our research and that of others to apply these chemistries to the development of both devises for implantation or as remediation systems.

At times it may appear that we are concentrating on one field or the other. Remember however that we as chemists consider both a device implanted in the body and the development of a system to remediate air or water as organs and therefore work on similar engineering principles.

References

1 *The ICI Polyurethane Book*, Wood, G. John Wiley & Sons, Ltd, Chichester, 1987.
2 *Dow Polyurethanes Flexible Foams*, Ed. Herrington, R. and Hock, K., Dow Chemical Company, Midland, p. 2.5.
3 On the Division of Space with Minimum Partitional Area, Kelvin, L. (Sir William Thomson), *Philosophical Magazine*, Vol. 24, No. 151, p. 503 (1887).
4 US Patent Nos. 6617014, 6991848 and 7048966, Dow, 1997.
5 Mechanism of cell detachment from temperature-modulated, hydrophilic-hydrophobic polymer surfaces, Okano, T., Yamada, N., Okuharo, M., Sakai, H., and Sakurai, Y., *Biomaterials* Vol 16, Issue 4, March 1995, pp. 297–303.
6 Impact of polymer hydrophilicity on biocompatibility, Ayala, H.Y., Sullivan, C., Wong, J., Jennifer, W., David, L., Chen, M., *et al.*, *Journal of Biomedical Materials Research*, 2009.
7 Biodegradable polyurethanes for implants. II. In vitro degradation and calcification of materials from poly(epsilon-caprolactone)-poly(ethylene oxide) diols and various chain extenders, Gorna, K. and Gogolewski, S., *Journal of Biomedical Materials Research* 60, 592, 2002.
8 The capsule quality of saline-filled smooth silicone, texture silicone, and polyurethane implants in rabbits: a long-term study, Bucky, L.P., Ehrlich, H.P., Sohoni, S., and May, J., Jr., *Plastic and Reconstructive Surgery* 93, 1123, 1994.
9 An assessment of 2,4-TDA formation from Surgitek polyurethane foam under stimulated physiological conditions, Szycher, M. and Siciliano, A., *Journal of Biomaterials Applications* 5, 323, 1991.
10 Changes in macrophage function and morphology due to biomedical polyurethane surfaces undergoing biodegradation, Matheson, L.A., Santerre, J.P., and Labow, R.S., *Journal of Cellular Physiology* 199, 8, 2004.
11 Development of a YIGSR-peptide-modified polyurethaneurea to enhance endothelialization, Jun, H.-W. and West, J., *Journal of Biomaterials Science Polymer Edition*, 15(1), 73–94, 2004.
12 Development of degradable polyesterurethanes for medical applications: in vitro and in vivo evaluations, Saad, B., Hirt, T.D., Welti, M., Uhlschmid, G.K., Neuenschwander, P., and Suter, U.W., *Journal of Biomedical Materials Research* 36, 65, 1997.

13 Identification of biodegradation products formed by L- phenylalanine based segmented polyurethaneureas, Elliott, S.L., Fromstein, J.D., Santerre, J.P., and Woodhouse, K.A., *Journal of Biomaterials Science Polymer Edition* 13, 691, 2002.

14 In vitro degradation of novel medical biodegradable aliphatic polyurethanes based on ϵ-caprolactone and Pluronics® with various hydrophilicities, Katarzyna Gorna, K. and Sylwester Gogolewski, S., *Polymer Degradation and Stability* 75(1), 113–122, 2002.

15 Recent developments in biodegradable synthetic polymers. Gunatillake, P., Mayadunne, R., and Adhikari, R., *Biotechnology Annual Review* 12, 301–47, 2006.

16 Surface modification of poly(ether urethane urea) with modified dehydroepiandrosterone for improved in vivo biostability, Christenson, E.M., Wiggins, M.J., Anderson, J.M., and Hiltner, A., *Journal of Biomedical Materials Research* 73A, 108–115, 2005.

17 New isocyanates from amino acids, Hettrich, W. and Becker, R., *Polymer*, 38(10), 2437–2445, May 1997.

18 Biodegradable block copolymers as injectable drug-delivery systems, Jeong, B., Bae, Y.H., Lee, D.S., and Kim, S.W., *Nature* 388, 860–862, August 28, 1997.

2

Laboratory Practice

Introduction

We thought it would be useful to include a chapter on how one would approach a problem when polymers are considered a solution. You will see many examples when we discuss how polymers are used and adapted to meet product needs. Whether it is an implantable system to repair bone defects or remove odors from an industrial plant or municipal waste treatment facility, polymers are increasingly considered an appropriate technology. Of course, our purpose here is to show how one can apply polyurethanes, not as off-the-shelf solutions, but rather as a set of tools to be applied. This shifts the emphasis to the laboratory to make best use of the tools the polymer offers. We will begin with commercial products and prepolymers and show how they are used in the laboratory. This has an educational aspect as you will become familiar with commercial products and get accustomed to the techniques of foaming and elastomer preparation. Depending on the nature of your project, there is likely a product or prepolymer that meets your needs. This is certainly the case when the physical properties of the resultant polymer are the primary requirement. If these materials meet the requirements of your project, scale-up is a simple matter. If, however, the requirements of your research includes an emphasis on the chemistry and that chemistry is not possible in commercial products, your focus is then shifted first to commercial prepolymers and then to synthesis of unique molecules. Having said that, the next step in your exploration would be in commercial prepolymers. Recognizing that commercial polyurethanes were developed to meet commercial needs, seat cushions, for instance, prepolymers are designed to have a much broader range of applications. We will give a range of products available to you at moderate cost. We will talk about some remarkable research using a commercial prepolymer. For instance, a commercial

Polyurethane Immobilization of Cells and Biomolecules: Medical and Environmental Applications, First Edition. T. Thomson.

prepolymer was loaded with a ceramic material and fired to yield ceramic foam. We will discuss this later and reference it then. In our work, we used an elastomer formulation to create a vacuum-sealed actuator, again with a commercial prepolymer.

We recognize however that your use of a polyurethane must have certain unique requirements, both chemically and architecturally. Among those chemical features are biodegradability, compatibility, and surface-active agents. Accordingly, your research will take you to a direction that requires a set of tools not included in commercial products. We will discuss those in detail in subsequent chapters and give examples from our laboratory and that of other on how to address those limitations.

Prepolymers

In the next chapter we will be describing scaffolds. As part of that discussion, we will give examples of how researchers have approached the subject from the perspective of thermoplastic polymers like polyethylene and foams of several types. In order to build what is generally accepted to be a suitable architecture, researchers using thermoplastics and ceramics have used a technique we call a "reverse lost wax process." By this technique a particulate is added to the polymer, and after curing, the particulate is removed by leaching with a solvent or heating. The result, of course is the required porous structure.

Without commenting on the advantages of this compared to a direct foam process, if this technique has some advantages, it is appropriate that we describe how one would use the method with polyurethane chemistry. There are a number of formulators who produce prepolymer systems to produce an appropriate material, called elastomers. Such a procedure is described generically.

Preparation of an Elastomer

Each manufacturer has a set of procedures that they recommend. The following is a generalized procedure using a two-part system. For clarity we will use the most common convention with "Part A" being the isocyanate and "Part B" the polyol, catalyst, surfactant, and so on.

While a fume hood is always preferred, for ventilation and isolation of the components, operating in an open area with a good air circulation or well-ventilated area can be done safely. Normal precautions of rubber gloves, long sleeves, and protective eyeglasses to avoid skin/eye contact are advised. The material is relatively safe for its kind, but still can cause some skin irritation. The MSDS of the materials should be available and read by the operators.

A mixing container can be a regular polymer paint bucket, large empty coffee can, lab beakers, or any other round-shaped dry and clean container. Disposable containers are always preferred for mixing.

If the product is to be molded, you will need to use a mold release for ease of unmolding. Silicone mold release systems are preferred. They should be applied just before pouring. Do not use water-based mold release agents.

Some materials are very thick (high viscosity) and difficult to handle. To make it easy to handle and mix well, you can heat the material within the range of 100–180°F (38–82°C). Refer to the product literature for instructions.

With all the materials available, set aside two containers, one for mixing the components and the other for curing. Mixing can be difficult and so we always advise that you mix the best you can and then pour the liquid into another container. Begin by applying the mold release in the container for curing.

After determining the relative and total amounts of the parts, following the manufacturer recommendations or your experience, pour the necessary amount of one of the parts into the container as you weigh. Which component should be added first is not always clear. Adding the higher viscosity component is a general rule, but ultimately the order should be governed by which achieves the best mixing. Close the container after pouring. Remember, the isocyanate is very sensitive to water, even humidity.

Pour the second part into the container as you weigh. Again close the container and begin mixing. Mixing by hand is possible, but using a handheld powered mixer is preferred. In that case mix at a lower speed using a paint-style blade. In either case, avoid mixing air into the system.

When you are satisfied with the quality of the mix, pour it into the mold container. If necessary, place the container in a vacuum chamber. Pull a vacuum until you see most of the bubbles are gone.

The material cures for several days. If you need to conduct physical test sooner, you can post cure at 180°F for 12–16 h. In any case, after 12 h or so, the material may be solid but still soft. Remove it from the mold slowly and carefully. Physical testing should be delayed for a full 24 h.

If you are not using disposable containers, you can use MEK, acetone, isopropanol, or other industrial solvents to clean them.

As we said, isocyanate is highly reactive to water and moisture. When you are done with mixing, the material should be blanketed with nitrogen gas or dry air, and the lid has to be closed tightly for storage.

You can use polyurethane grade colorants/pigments to add colors to the material. Particulate dispersion-type colorants are commonly used with this type of polyurethane. They are typically dispersed into glycol ether to form what is called a color concentrate. These are added to the polyol part at concentration of 0.1–2% by weight without affecting the properties of the polyurethane.

Finally, fillers can be added. Commonly used fillers are clay, silica, wollastonite, or glass.

Preparation of Foam

The procedure to make a foam from a prepolymer depends on the type of foam required. The major differentiating factor is whether the product is to be hydrophilic or hydrophobic. Both types are used in scaffold development, but the processes are dramatically different. In this section we will focus on hydrophobic prepolymers. We will then devote a significant amount of time on hydrophilics.

Hydrophobic Foams

While we don't make any recommendations to a particular company, we have worked with many of them. They are the experts in this field, and while your research may take you in a completely new direction, you would be wise to use their expertise as a foundation. We will use data from Northstar Polymers (Northstar Polymers, LLC, Minneapolis, MN, USA), again not as a recommendation, but as a representative of the industry. The polyurethane foams that interest us are open-cell foams. While void volume and interconnectedness are our primary focus, these formulations are designed with other properties in mind. Nevertheless, commercial open-cell flexible foam is a good starting point.

The following foam formulations are designed to make flexible foam parts/sheets/dies/blocks by hand-mixing or by metered dispensing equipment. The components are stable liquid at room temperature. These properties are ideal in small-scale productions for custom foam applications. While they were formulated to make cushions and packing products, as we said, they are a good starting point. Generally, lower density grade foams are softer than higher density foams.

The following two-part products are available commercially (Table 2.1).

We will explain this chart later when we discuss custom prepolymers. You may want to refer to these numbers at some point. Parts A and B have certain reactive characteristics that define the ratios in order to produce a stable foam (Table 2.2).

Table 2.1 Commercially available open-cell foam producing prepolymers.

Density (kg/m^3)	Approximate void volume (%)
0.62	93
0.83	90
0.97	89
1.1	87
1.2	86

Source: Adapted from Northstar Polymers, LLC, Minneapolis, MN, USA.

Table 2.2 Reactivity of parts A and B.

	Prepolymer (part A)	Polyol (part B)
Specific gravity	1.183	1.024
Equivalent weight	183	274
%NCO	23	

Source: Adapted from Northstar Polymers, LLC, Minneapolis, MN, USA.

Table 2.3 Typical values from a commercial prepolymer.

Property tested	Result
Foam density (free rise)	0.70 kg/m^3
Compression deflection (25%)	11 kPa
Compression deflection (50%)	27 kPa
Rebound	28%
Tensile strength	260 kPa
Tensile elongation	121%

Source: Adapted from Northstar Polymers, LLC, Minneapolis, MN, USA.

Typical values of a fully cured foam from one of the previously mentioned formulations are shown in Table 2.3.

The following procedures are used to make these foams. This is based on an MDI prepolymer and polyether polyol. The components of these formulas are stable liquids at room temperature. The flexible foam formulations come with different densities and firmness to meet the application requirements. This procedure is used to make foam by hand-mixing or by metered dispensing equipment.

As with the elastomers, operate in a well-ventilated area or large open area and wear rubber gloves, long sleeves, and protective eyeglasses to avoid skin/eye contact. Read the Material Safety Data Sheet for details on the safety and handling.

Freezing can be a problem with the isocyanates. This causes phase separation within the components. If you are concerned that part A might be frozen, heat it to 140–160°F and agitate the content. After agitating the component, keep them at a room temperature above 70°F.

If you are making a free-rise foam, you can calculate the total mass of material from the expected final density. If you will be using a closed mold, add 5% to the free-rise mass. The expansion pressure of the foam fills the mold. The mold has to have a capacity to retain the internal pressure.

The air trapped in the mold needs to be vented. Very small holes need to be drilled to let the trapped air escape. The placement of those holes need to be situated so they do not affect the purpose of the product. Careful calculation of the amount of material is required.

The mold material can be metal, plastic, or an elastomeric material. Mold surface must be without visible flaws. Even then, multiple use of a mold will eventually cause it to foul, making it unusable. The pressures will eventually force material into microscopic pores. In our experience, polymer molds are more economic. Having said that, metal molds with their higher heat capacity can be used to some advantage.

The process is virtually the same as with the elastomer process described earlier. Pour the correct ratio of part A and part B into a mixing cup. Mix well with a steel or plastic stirrer or stick for about 10 s. Agitate vigorously and thoroughly.

MDI-based prepolymers react at a different rate than a TDI product. This doesn't matter so much with elastomers, but with foam, timing is critical. Inasmuch as we are mostly concerned with foam architecture, mixing past "cream time" destroys the natural structure. Refer to the description of the curing process presented in the hydrophilics section to follow. Edge effects are obvious and this needs to be accounted for when analyzing the product. This is not to say that hand-mixed foams are without value. It must be kept in mind that careful analyses should be done on meter/mix devices in which the blending and deposition are completed before cream time. The larger the sample size, the more important a dedicated meter/mix device is important. We will mention a few devices we have used after we discuss hydrophilic prepolymers.

After mixing immediately pour the mixture into the mold. Cure the foam in the mold for at least 20–30 min. Check the strength and tackiness of the surface. Tackiness is a symptom of an incomplete cure. Trying to remove the foam while it is tacky invariably destroys the product. Be aware that the lack of tackiness does not mean the foam is fully cleared. As with the elastomer, full cure can take 24 h at room temperature. Once you are comfortable that the foam is cured enough to handle it, remove it from the mold. Depending on the size of the sample, a band saw is used to remove the top, bottom, and side. What is left is suitable for physical analysis. The first test is to compress the foam with your hand to determine if the foam has an open-cell structure. If it feels like a balloon, the foam cells are not opened. If it compresses easily and bounces back immediately, the foam is open-celled.

Hydrophilic Foams

It can be said that hydrophilic prepolymers follow the same reaction scheme as we illustrated earlier. While part A is the isocyanate phase, part B is not a polyol plus catalysts, and so on. Part B is water plus a small percentage of a surfactant.

The surfactants are added to improve emulsification. In fact, the surfactants, while not changing the chemistry, have a major effect on the architecture of the resultant foam. Without a surfactant, a closed cell foam results. Depending on the surfactant, the material can be extremely small cell (<100 μM) or a foam structure that approximates a sea sponge. It is the water however that is the reactant that both cures and causes the mass to foam. Review the first chapter for an elaboration of the water reaction. While all foamed polyurethanes include water in the formulation, the volume of water used to produce hydrophilic polyurethane foam makes it unique. Water can be as much as three times the mass of the isocyanate phase as opposed to a few percent for a hydrophobic. If it weren't for the fact that the water reacts with the isocyanate, the hydrophilic prepolymer would dissolve. Before we get to the purpose of using a hydrophilic polyurethane prepolymer, some other reactions are worth noting. While the predominance of products is as foam, elastomers can be made. Difunctional amines or carboxylic acids react quickly with the isocyanate to produce a hydrophilic elastomers. The reaction is called chain extension inasmuch as the molecular weight increases. We will talk about these reactions and others in the next section when we discuss custom prepolymers.

Returning to conventional uses, the water has three purposes. Firstly, it is a critical part of the reaction. All foam formulations include water, even hydrophobics. As we explained in the chemistry chapter, it reacts with the isocyanate to produce CO_2 that expands the mass. As important as it is, a very small amount of the water that is used reacts. When it does, however, in addition to the foaming process, it creates an amine end group. This is the second purpose of the water. It is the amine that reacts with other isocyanate end groups to begin the process of building molecular weight.

Thirdly, the water functions as a heat sink, helping to control the exotherm. This aspect will be important when we discuss the immobilization of enzymes or other proteins sensitive to denaturation.

One might add a fourth function in that it creates a temporary matrix within which the foam is stabilized until the polymer can support itself.

We thought it would be useful to give a narrative of the foaming and curing process. There are points during the reactions that offer the researcher options that may be useful in building an appropriate scaffold. Variables that we will discuss include the surfactant and temperature. This description is based on Hypol® 2000 or equivalent and a water ratio of 2:1 (water to prepolymer). We will use Pluronic L62 (BASF Corp.) about 0.05% based on the water. It produces a small open-cell foam. It is not a good filter material as it collapses under moderate pressure. However, it is an excellent chronic wound dressing. It is very absorbent and absorbs liquids quickly (wicking, discussed later). As an aside, the cells can be opened up slightly by adding a little Pluronic F68. It doesn't stiffen the foam but it increases airflow, discussed in the analytical section of this chapter.

In this scenario, we weigh out an amount of prepolymer into our mixing cup and then weight out the water phase containing the surfactant (the aqueous). Pigments and fillers can be added but they do not effect the reaction. The aqueous is then added to the prepolymer and the clock starts. Using a handheld drill fitted with a flat paint mixer, we immediately begin blending. Foaming begins almost immediately. As the CO_2 is released, amine end groups are foamed and immediately act as chain extender, and the mass begins to build molecular weight (i.e., get stronger). This process continues, building strength, while the internal pressure expands the foam. After a few seconds (ca. 10–15), the strength of the polymer is high enough to begin to restrain the CO_2 pressure, and the expansion slows. Shortly thereafter is the point called "cream time" or as we refer to it as the gel point. It is called cream time because in a continuous pore system, if one continues to add the emulsion, it appears to separate from the foam and it has the appearance of cream separating from milk.

If you were to disturb the foam at this point, the natural structure that is being built would be disrupted. Interesting however is if you were adding pigments to the aqueous, a visible marbling effect is created.

If you are mixing with anything other than a commercial meter/mix machine (discussed shortly), fully blending the liquids before cream time is nearly impossible. Thus one has to realize that hand-mixed foams are close, but not the same as machine-made foams. We have tried to make a mixer that mitigates this problem without success. Because of this, one must be careful to extrapolate hand-mixed samples to commercial or clinical grade.

Cream time is an event in the curing process, but this clearly does not mean that the reaction is done. If you were to touch the foam at that point, you would see that it is very soft and sticky. Carbon dioxide is still evolving, but rather than increasing the volume, it begins to break the windows between the pores. Anthropomorphically, the CO_2 is trying to escape.

It is appropriate at this time to introduce the effects of temperature. As we mentioned there are two reactions occurring: gas generation and increasing molecular weight. They both increase with temperature, but not at the same rate. Cold temperatures favor polymerization as opposed to gas generation. High temperatures favor gas generation as opposed to polymerization. Both reactions produce heat, but the temperature of the reactant must be controlled to produce a uniform process. Case in point is the immobilization of enzymes and of biological systems. Care must be taken to prevent denaturation. A normal process to make foam might be at 45°C, but that is too high given the exotherm. Thus you will see that researches have cooled the reactants to about 4°C. This has a dramatic effect on the density and pore structure. For the most part, a closed cell foam is produced. When we discuss immobilization, you will see that it is necessary to cut the foam into pieces to increase surface area unless certain procedures are used to open the pore structure.

For commercial processes, higher temperatures increase production rates, but the temperature must be moderated to produce the appropriate foam structure. Two examples illustrate. We produced a wound dressing for a major healthcare company that we immodestly speak of as the ultimate in chronic wound care technology. The formulation was straightforward, but the temperature was controlled so precisely that we could control the amount of open-celledness. We described it as a matrix of water pitchers. Thus one could "fill" the foam with water and then hold it up by a corner and only a small amount of water would drain. A typical foam would lose about 75% of the water. Obviously the pitcher foam would be inappropriate for cell spreading.

Compare this to a consumer product we worked on. While cold temperatures lead to a high density closed cell foam, a company wanted to make a physically weak, high density foam, two factors that work against one another. Fillers were added but the trick was to run the process at high temperatures. In fact, the temperatures were so high that the CO_2 developed enough pressure (relative to the polymerization) to escaped the mass, and thus the foam collapsed.

Returning to the process scenario, the process continues, and while you don't see any physical effects, gas is still being created but at a slower rate. The polymerization is continuing. The next point of interest is what is called tack-free time. As the name implies, after this point, the foam can be touched and while it is still weak, it is not sticky. Polyurethane pressure sensitive adhesives are produced by stopping the reaction before this point. Related to this is a popular adhesive that is actually an MDI prepolymer. You are asked to moisten the materials to be bonded, apply the adhesive, and then use clamps to hold them in place. When the adhesive comes into contact with the wet surfaces, enough CO_2 is produced to force the adhesive into cracks and pores before the final cure. Recall our caution in using pressurized molds. Eventually they foul as the polymer invades any imperfection in the mold. You can test the expansion on these adhesives by mixing a little water with one of these adhesives or even leave the top off the container to expose it to humidity. After a little time you will see that it forms a hard but brittle foam.

Again back to the story, tack-free time the foam is still not strong enough to remove it from the mixing cup. That will take another 10 min or so. Even then, it will take another 24 h or so to complete the reaction.

We will show how these prepolymers are made, but we think it is wise to begin your research using commercial products. These prepolymers were designed to produce foams that will absorb several times its weight in water. We will show how this can be increased or decreased, but these materials provide a good starting point and will be an economically useful way to learn the techniques of foam production. In addition it will give you access to knowledgeable technical service people. In that regard, there are many hydrophilic prepolymers available. The commercial products invariably use polyethylene glycol and a small percentage of a triol. TDI and MDI versions are available. We have used most if not all of the prepolymers produced in the United States and Europe and found them to be roughly equivalent.

The techniques that have been developed to control all the aspects of commercial hydrophilic polyurethanes are beyond the scope of this book. The formulations, methods, economics, and processes specific to hydrophilic polyurethanes are described more completely in an earlier book [1]. It is appropriate to focus on those issues that concern the development of an environmental or medical scaffold. In that text the use of surfactants are discussed in some detail. The use of the foam on the treatment of chronic wounds is the primary focus, but in doing so the process to make and then test the material applies to many applications, and especially the subject of this book.

Again the primary medical use of hydrophilic foam focuses on its ability to absorb liquids. While it is used as a non-soil growing medium for plants and as a cosmetic applicator sponge and dozens of other applications, our focus here is on advanced medical and environmental devices. In that role while being absorbent, it has the generally considered desirable property of being compatible with water. It is in every sense of the word a hydrogel, even in its foam state. You saw from the discussion earlier and will see in the chapters on scaffolds and immobilization. Many researchers have used hydrophilic prepolymers to develop devices from dressings to incontinence devices. Our involvement started when wounds were still treated by dry gauze and relying on normal scabbing. Hydrocolloid dressing moved the technology toward wet dressing but the evolution took a great step forward by Smith and Nephew in England and Ferris in the United States with dressings based on hydrophilic polyurethane.

This technology, in a sense, recognizes that the body accepts water-based systems as being natural. Those of you who work in this arena know better than I the effect of hydrophobic surfaces. In the discussions that follow, we will stress the need for an appropriate scaffold. It requires certain architectural aspects that we will discuss in some detail. The attention then inevitably shifts to the chemistry of that surface. Biocompatibility and immobilized biomolecules are essential but so is a surface that includes components appropriate to an extracellular matrix. Growth factors and similar functional components will become critical. In that regard, we will of course discuss the surface and modifications thereof, but one should also be aware of and take advantage of the ability to absorb soluble materials and then deliver them. Nutrients and salts can diffuse out of the polyurethane in a controlled manner. We have studied it as an electrophoresis reservoir system.

To complete the discussion of commercial prepolymers, we wanted to focus our attention on the practice of making a hydrophilic foam or elastomer, relative to its use as a scaffold for cells or biomolecules. In this regard, the ability to pass air or water through it is important. For an elastomer, you know by now that the enemy is water. If you are to make a film of this material, you may choose a commercial polyfunctional amine. They contain primary amino groups attached to the end of a polyether backbone, which is normally based on either propylene oxide (PO), ethylene oxide (EO), or mixed PO/EO. While again not as a recommendation over other suppliers, we have used the Jeffamine® family of functional amines. In their catalog are EO-based di- and triamines. The reaction with a hydrophilic prepolymer is very fast and so a diluent is necessary (acetone or toluene is appropriate). Dissolve the prepolymer in the solvent and then add the amine slowly with stirring.

If you were to make a foam, one would want it to have certain characteristics. Open cell is of course the first of those, but as we will discuss in the next chapter, void volume, interconnected cells, physical strength, and other factor are part of the package of a successful scaffold. Unfortunately they cannot all be developed in a conventional hydrophilic foam. It is a hydrogel however and as such is fulfills one of the two primary aspects required of a successful scaffold. Those aspects as we will mention several times are chemistry and architecture. It is the chemistry that makes it interesting. As we will show, apart from its hydrophilic nature, the ease at which it immobilizes proteins is impressive. Its ability to absorb biological fluids like plasma and potentially release them makes it somewhat unique in the polymer business.

It is in the category of strength of materials that commercial hydrophilic polyurethanes need help. As one might suspect the mere fact that it is 70% water when hydrated, limits at least tensile properties to the category of "weak." This is addressed but it is certainly an area of concern for scaffold development.

If you will allow us, we often tell a story that brings this into perspective. When God designed us, he/she realized that while the functional parts of the body are competently served by hydrogels, but the structure was not. Thus the design had to be based on a balance of structural bone and functional hydrogel.

In our work, we used reticulated foam as the bones, but in fact as long as you recognize the need for structural support, almost any material can be used. In the later sections, hydroxyapatite and clay are suggested.

Thus the surface of a scaffold must be able to accommodate these possibilities. You know by now that we think a hydrophilic polyurethane will become an important "gateway" to the eventual solution. Current hydrophilics are built around polyethylene glycol and an aromatic isocyanate. You will see in the following chapter that this may evolve into aliphatic polyurethanes with copolymers of ethylene glycols and polypeptides. It is said (and we have no reference) that a polymer that absorbs greater than 30% of its weight in water is potentially biocompatible. This gives us lots of room to develop an appropriate polyol, polymerized and foamed by an isocyanate that meets a biochemical need.

In that regard and by this, we complete our discussion of hydrophilic prepolymers; while we typically think of them as the route to a foam, you might also consider it as a surface treatment. The device that you envision will have several aspects including strength of material and chemical compatibility. High water contact and strength are mutually exclusive. A logical solution is to make a composite material the components of which serve one of the purposes. Consider wood; it has high strength fibers imbedded in a lignin adhesive. Carbon fibers while having high tensile strength are of little use unless oriented in an epoxy matrix.

With that let us move away from commercial materials and give an overview of what I believe to be future prepolymers.

Custom Prepolymers, Foams, and Scaffolds

We have discussed two processes by which flexible polyurethane foams are made. The first is the one-shot process by which the ingredients are slammed together by use of an impingement mixer and deposited in to a mold or onto a moving conveyor. This is an efficient way to manufacture large quantities of foam. In fact it made polyurethane the success that it is. It is however to our knowledge limited to large production and therefore inappropriate to the research that needs to be done. Ultimately, this may change with the development in environmental remediation, but for now the other methods, prepolymers, are the most appropriate approach. It may be interesting to note that hydrophobic polyurethanes are produced in the billions of pounds, compared with hydrophilics in millions.

By way of review, prepolymers are the result of the reaction of a diisocyanate with a polyether or polyester. For purposes of clarity, and in the sense that the scope of our research includes other reactants, we will use the most common reactant, the polyol, as shorthand for any polymer the end group of which reacts with an isocyanate. This includes, but is not limited to, amines and carboxylic acids. In subsequent chapters we will discuss polypeptide with their amine functionality as an alternative to polyethylene glycol, for instance. So keep in mind that we are not limiting the chemistry to polyols even though I use that term.

As we have discussed, the reactions take place at the end groups. Figure 2.1 represents the reaction to make a prepolymer. Note that it takes two molecules of the isocyanate to react with, in this case, a polyethylene glycol. The result is described as an isocyanate-capped polyol.

Our discussions have used hydrophilic polyols, but of course there are hydrophobic prepolymer as we explained earlier in this chapter. When those are used, only a small amount of water is used (ca. 5% by weight), but the bulk of the polyol is a hydrophobe. In that case, the reaction rate is slow and so

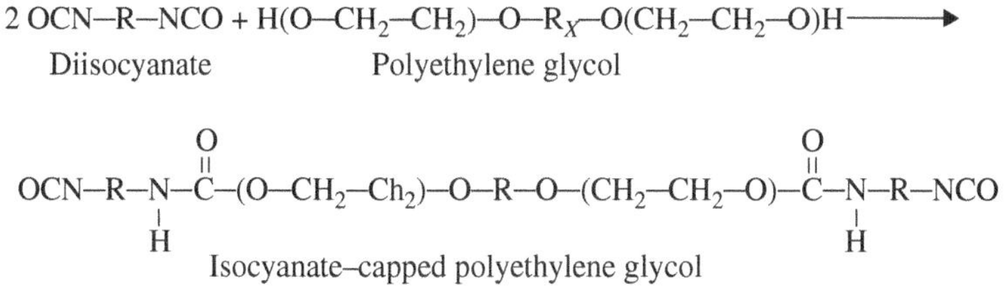

$$2\ OCN\text{—}R\text{—}NCO + H(O\text{—}CH_2\text{—}CH_2)\text{—}O\text{—}R_X\text{—}O(CH_2\text{—}CH_2\text{—}O)H \longrightarrow$$

Diisocyanate Polyethylene glycol

$$OCN\text{—}R\text{—}\underset{\underset{H}{|}}{N}\text{—}\overset{\overset{O}{||}}{C}\text{—}(O\text{—}CH_2\text{—}Ch_2)\text{—}O\text{—}R\text{—}O\text{—}(CH_2\text{—}CH_2\text{—}O)\text{—}\overset{\overset{O}{||}}{C}\text{—}\underset{\underset{H}{|}}{N}\text{—}R\text{—}NCO$$

Isocyanate–capped polyethylene glycol

Figure 2.1 The prepolymer reaction.

catalysts are used. Note that mentioned earlier but in both cases, surfactants are used to help with emulsification. Emulsification effects pore size and open-celledness. No catalysts are needed with the hydrophilic prepolymers as the reaction of water is nearly instantaneous as evidence by the immediate foaming of the mixture.

The production of a quality polyurethane foam depends on several factors including mechanical and chemical. While not shown in the last figure, branching adds an additional reactive "end" groups to the molecule. This can be intentional by the addition of a triol, but more often than not it is due to imperfections in the polyol.

Many times rather than using the molar ratios, researchers and production technicians use equivalent weights (EWs) to determine the ratios of reactants. You will remember that the EW is the molecular weight divided by the reactive groups. Thus a 1000 MW diol has an EW of 500. According to the rule of equivalent weights, an equivalent weight (EW) of polyol will react with an EW of a diisocyanate (MW/2).

In the industry, the amount of hydroxyl is given as the hydroxyl number. In fact, while they are called diol (2 "–OH"), they are in have somewhat more –OH character. Thus a procedure was developed to measure the hydroxyl. The actual EW is then given by

$$EW = \frac{56.1 \times 1000}{OH\ number}$$

The method to determine hydroxyl numbers are in ASTM D4274-94d.

The difference is not great, and in the context of laboratory preparations, they are negligible. Typically, we use an excess isocyanate anyway. Still further, inasmuch as we know, the strength of the resultant foam is determined in part by the amount of functionality, it is common to add a small amount of triol. In the case of commercial hydrophilic polyurethanes, this can be as much as 5%. Using this figure, the amount of isocyanate is given by the following calculations:

$$\text{Mole of isocyanate} = (0.95 \times 2.0) + (0.05 \times 3)$$

Thus 2.02 moles of isocyanate are required for each mole of polyol. In practice, an additional amount of isocyanate is added to increase the reaction rate. A prepolymer with an excess isocyanate also seems to foam more uniformly and certainly more quickly.

With regard to the cross-linking effect, as we discussed in the chemistry chapter, the functionality plays a critical role in the mechanical properties of the resultant foam. A functionality of 3.0 makes a hard, brittle foam, regardless of other factors.

Continuing with functionality, if the calculated value shown earlier is used the molecular weight would be maximized but so would the viscosity. This could make it difficult to convert into a uniform foam. This is another reason to add an excess of isocyanate. Also care must be taken if the cross-linking agent is an amine as this increases the reaction rate and also interferes with the CO_2 to amine process of building molecular weight. Multifunctional amines can be used but with this cautionary comment. We have done quite a bit of research on the use of trifunctional amines under that trade name of Jeffamine.

While this is certainly a variable to be investigated, reaction temperatures for TDI and MDI prepolymers between 65 and 100°C are typical. For other isocyanates, like isopherone diisocyanates, the temperatures for a reasonable reaction time are more in the range of 125°C. Above 100°C, uncontrolled cross-linking reactions can occur. The reaction time is not a precise point. Typically viscosity is used to indicate the progress of a reaction. By way of example, a typical time/viscosity relationship is shown here (Figure 2.2).

Other factors affect viscosity including acidity of the isocyanate. As part of your research design, this should be included in your agreement with your supplier. In all cases, while we are discussing it, one should develop a certification protocol for all your raw materials. Along with the acidity and chlorides of the isocyanate are the water content and the hydroxyl number (or other functionality) of the polyol.

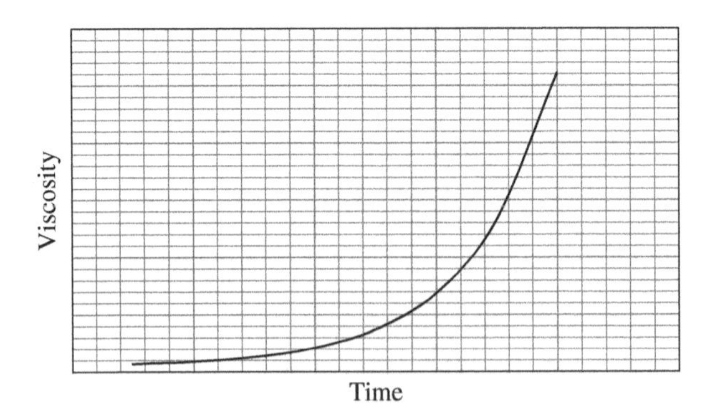

Figure 2.2 Typical viscosity increases as a function of time.

In summary, we have tried to develop an overview that includes the ratio of isocyanate to polyol and other factors that control the quality of the resultant prepolymer. Our focus has been on hydrophilics for this discussion, but a review of the procedures to make a hydrophobic foam from commercial prepolymers would show that the reactions, if not the procedures and formulations, are different.

Examples

With that as an introduction, we wanted to give an example of a process to make a prepolymer. In this example we will use a process that mirrors that of a hydrophilic polyurethane prepolymer, but for reasons of confidentiality, it is best that we use an example from the literature. In this case the prepolymer and foam development were described in Saunders *et al.* based on work at DuPont in the 1950s [2]. In this study, various amounts of propylene glycol (MW = 2000) and several triols and tetols were reacted with the isocyanate to produce prepolymers. The following figure details the formulations. The references were not given in Saunders, and the paper is an internal report of E, I DuPont de Nemours & Co. We can assume however that the reaction was at 125°C for several hours (Table 2.4).

The results of the reactions, while not dramatic, give some guidance in building our prepolymers. It can be taken for granted that changes of this magnitude could be significant (Table 2.5).

Somewhat closer to the interests of our medical readers but still on the subject of prepolymer preparation, 10 polyurethanes were prepared to investigate fibrinogen adhesion to polyurethane surfaces [3]. Of the 10, one was a commercial product, Biospan® SPU (Polymer Technology Group, Berkley, CA). Of the 10, we will examine 6 of them, 1 of which is the commercial product. They are labeled in the paper as PU5 through PU10 (Table 2.6).

Table 2.4 Formulation for prepolymer preparation.

Formulary	C-1	C-2	C-3
Polypropylene glycol	75	100	100
Triol 1 (MW = 2500)	25	—	—
Triol 2 (MW = 134)	—	3	—
Tetrol (MW = 294)	—	—	0.5
Water	0.15	0.15	0.15
Isocyanate (based on equivalent weight)	1.05	1.05	1.05

Source: Hettrich [10]. Reproduced with permission of Elsevier.

Table 2.5 Properties of the resultant foam.

Analysis	C-1	C-2	C-3
Viscosity (30°C in cps.)	5300	18 000	6300
% NCO	0.09	0.09	0.09
Gel strength during foaming	Fair	Fair	Good
Density (kg/m^3)	2.2	2.1	2.3
Tensile strength (kPa)	15	12	19
Compressive strength at 25% (kPa)	0.45	0.53	0.44

Source: Hettrich [10]. Reproduced with permission of Elsevier.

Table 2.6 Formulary of the Wu study.

Exp.	Polyol	Iso	Chain extender	Mole ratios polyol: iso: chain extender
PU-5	PTMO 2000	MDI	BATA	1:6:5
PU-6	PTMO 2000	MDI	EDA	Biospan
PU-7	PTMO 1000	MDI	DAHP	1:4:3
PU-8	PEG 2000/PEG1000	MDI	BD	0.5/0.5:4:3
PU-9	PEG 2000	MDI	BD	1:4:3
PU-10	Peg 1000	MDI	BD	1:4:4

Source: Jeong [11]. Reproduced with permission of Nature Publishing Group.

The process is described in Simonovsky [4]. The materials were produced by a multistep process beginning with an isocyanate prepolymer without chain extension. The polyol was added to a reactor (three-necked round bottom flask fitted with a stirrer and provisions to purge with nitrogen). The MDI was added and the reactor heated to 80°C with stirring for 2 h.

The resultant prepolymer was dissolved in dimethylacetamide as was the chain extender. The chain extender solution was added to the prepolymer with mixing. It was then heated to 80°C with a catalyst (dibutyltin dilaurate). Stirring continued for 15 min. Small quantities of MDI were added to achieve stoichiometry. According to the authors, "This gradual approach to stoichiometry was used to improve process control to produce reproducible, high molecular weight PEUs and control the viscous properties of their concentrated solutions."

Recognize that these are elastomers, a logical choice for surface analysis. The polymers were dissolved in *N,N*-dimethylacetamide and cast onto glass plates.

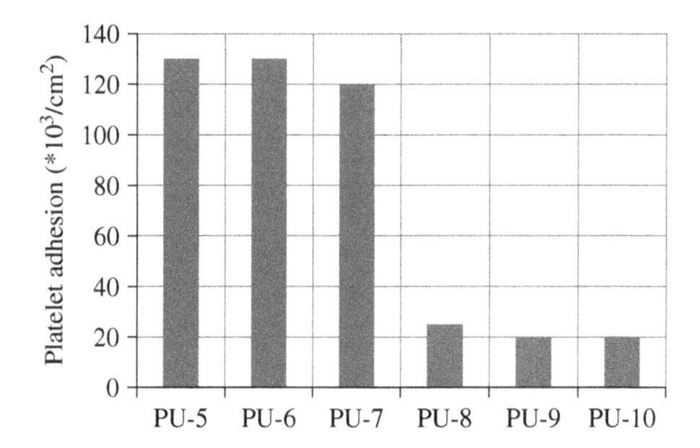

Figure 2.3 Effect of chemistry on platelet adhesion. *Source:* Adapted from Northstar Polymers, LLC, Minneapolis, MN, USA.

The polymers were exposed to the samples of protein (with a ^{125}I tracer) and incubated for 1.5 h at 37°C. Unbound proteins were washed away. The bound proteins were measured using a gamma counter. The results are shown here (Figure 2.3).

In another study, we were asked to investigate whether a polyurethane hydrogel (not a foam) could be developed that would qualify as an injectable that would cure in place and repair the discs between the vertebrae. It was an interesting synthetic study in prepolymer development, but a failure. The complexities involved in the material itself were challenging enough coupled with the problems of implanting the polymers made the project made the idea ultimately unsuccessful. Nevertheless, the chemistry is interesting and so we include it here. First of all, two hydrogel-yielding prepolymers were to be produced. Their physical strength, most importantly compressive, had to be equivalent to the natural cushioning between the vertebrae. In addition the design was to approximate 90% equilibrium moisture.

An intervertebral disc is a large, round ligament. It is made up of two parts. The outer ring is the annulus, which is the strongest part and is responsible for connecting the vertebrae. The inner area is the nucleus pulposus, which is soft and has the consistency of crabmeat. The nucleus pulposus acts as the shock absorber for the spine. It is this part was the subject of our investigation. The target mechanical properties were developed from confined compression studies, but there is no consensus as to how and what those numbers should be. As a guide however we used Perie *et al.* [5]. Our approach was to make a hydrogel prepolymer that would, when reacted with water, achieve 90% equilibrium moisture. In order to meet the mechanical goals, we used an excessive amount of cross-linking. While conventional cross-links are hard and hydrophobic, we identified a polyol that we felt was appropriate. Its trade name is

Carpol 5171 (Carpenter Co., Chemicals Division, Richmond, VA, USA). It is an EO/PO-based triol of approximately 5000 MW. In addition we used Arcol® Polyol HS 100, a polyether polyol modified with a styrene-acrylonitrile (SAN) polymer. It is specifically designed for increasing loadbearing properties of high resilience slabstock foams. Lastly Pluronic L62 was used to complete the polyol package. An experimental grid was developed among the polyols to determine the relative effects. MDI was used as the isocyanate. Mole ratios were calculated from the hydroxyl number for each of the polyols. The prepolymers were produced by addition of the isocyanate to the polyols and heated to 65°C for 4 h.

The formulations and results are shown in Table 2.7.

Sample 1 and 2 never solidified and so were dropped from the test. Others cured nicely in water to form decent hydrogels. Their tensile properties are presented in Table 2.8.

The object of the choices of the polyols, however, was to look at compression (Table 2.9). In that regard, formulation 5 was considered a candidate for expanded testing.

To simulate compression confined by the annulus, the round test samples were wrapped with black plastic electrical-style tape.

We report the following data not because it is significant but to offer a suggestion as to how one would report the compressive characteristic of a material (Figure 2.4).

Table 2.7 Experimental grid of spinal disc study.

Exp. no.	Pluronic® L62	Carpol®5171	Arcol® HS-100	MW	%EO
1	100	0	0	2500	20
2	50	50	0	3750	45.5
3	10	90	0	4750	65.9
4	40	40	20	4200	36.4
5	90	10	0	2750	25.1
6	10	70	20	4950	51.7
7	70	30	0	3250	35.3
8	0	100	0	5000	71
9	0	0	100	6000	0
10	0	50	50	5500	35.5
11	0	10	90	5900	7.1
12	0	30	70	5700	21.3
13	50	20	30	4050	24.2

Table 2.8 Results of the spinal disc project.

Exp. no.	Tensile strength (kPa)	Elongation (%)
3	0.3	124
4	0.2	159
5	0.04	265
6	0.16	120
7	0.12	149
8	0.38	131
9	0.45	161
10	0.21	126
11	0.22	176
12	0.09	195
13	0.23	119

Table 2.9 Samples for compression tests.

Thickness (mm)	Diameter (mm)	Area (mm²)	Area (inch²)
22.7	37	1074.67	1.67

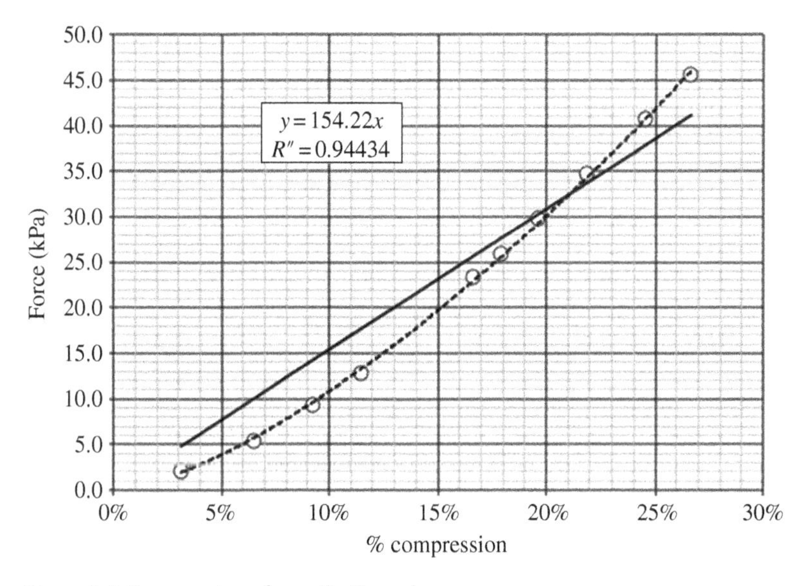

Figure 2.4 Compression of a synthetic nucleus.

This does not end the discussion of custom prepolymers. Numerous examples will be given in subsequent chapters. In some cases they illustrate alternative architecture, while in other cases they illustrate an immobilization technique. Occasionally they demonstrate both.

It is appropriate now to focus our attention on how one designs their polymer system.

Structure–Property Relationships

The purpose of this section is to give some guidance as to how one would affect the mechanical properties of your final product. Assuming you will be making your own prepolymer, the tools to develop appropriate architecture and strength of materials is in your hands. For the most part, this will be a qualitative description based on our experience in both hydrophilic and to a lesser degree, hydrophobics, which is where we begin.

We have used the concept of hard and soft segments to define the mechanical properties of the polymer. In this theory, the weight percent of so-called hard segments controls the stiffness of the polymer. In polyurethanes, isocyanates and cross-linking are considered hard segments. Polyethers and polyesters are soft segments, but polyethers are "softer" than polyesters. It is an oversimplification that MDI-based polyurethanes are stiffer than a TDI polymer in as much as the molecular weight, and therefore the mass % MDI is higher. For a given isocyanate, as the molecular weight of the polyol decreases, the stiffness increases. In this sense, stiffness is a synonym for compressive strength.

Regardless of the form of the polyurethane, elastomer or foam, the strength of a polymer is more often than not expressed as tensile strength. It is related to density, clearly, but it reflects in the inter- and intramolecular forces that hold the polymer together under stretching stress. A stretching force is applied to a sample and that force is measured as the sample gets longer. A plot of the force against the length of the sample is developed continuously until the sample breaks. The force at break, the elongation, and the slope of the curve are defined as the tensile properties of the sample.

Polyurethanes, like all polymers, follow the same rules when it comes to tensile strength. Molecular weight typically increases tensile strength due to the intramolecular tangling. Elongation also increases. Some of these effects are seen in our disc experiment earlier. Cross-linking increases the compressive strength.

Intermolecular forces also affect physical properties. Hydrogen bonding, van der Waals forces, and dipole moment cause the polymer to form crystallinity. The strength of these bonds is significantly lower in energy. Crystallinity is evident in all polyurethanes, and they affect the physical strength of the polymer in ways similar to intermolecular forces. In fact, they are closely related. Crystallinity causes brittleness of the polymer. These forces can be controlled by affecting changes in the process.

Cross-linking is also an important tool. As we have discussed, cross-linking is essential to foam making as it gives the reacting mass the ability to trap CO_2. In elastomers, it is an important factor as it controls tensile strength and elasticity.

At the molecular level, one can assume that the intramolecular bonds could have an effect on the mechanical properties. Each of the bonds has a cohesive energy that defines how much energy it takes to physically tear it apart. In the following table the cohesive energy of most of the important chemical bonds in a polyurethane is compared (Table 2.10).

Szycher [6] reported on the effect of isocyanates on the tensile strength of polyurethane elastomers using a 500 MW poly(oxytetramethylene) glycol. While we are dealing mostly with open-cell foam, control of the degree of openness also has an effect on tensile properties. Sanders [7] reported on the control of open-cell structure by the addition of a stannous catalyst. They show a maximum in their data that coincided with a maximum in tensile strength.

One can develop protocols to control the tensile properties of a polyurethane by adjusting the number of cross-links in the backbone. The unit weight of polymer divided by the molar number of cross-linking agent is referred to as the molecular weight per cross-link (M_c). It is a measure of the average molecular distance between cross-links. If a cross-linking agent such as glycerol or trimethylol propane is part of a prepolymer formulation, in our experience, it is just as easy to use the mass percent cross-linker.

Once the strength aspects of an anticipated device have been established, the design process is completed by the definition of a cell structure. If elastomers are the intent of the research, this section of the chapter will not be of interest, but if the project has flow-through aspects, in many ways the development of

Table 2.10 Molar cohesive energy of organic groups.

Group	Cohesive energy (kcal/mol)
–CH– (methylene)	0.68
–O– (ether)	1
–COO– (ester)	2.9
$–C_6H_4–$ (aromatic)	3.9
–CONH– (amide)	8.5
–OCONH– (urethane)	8.74

Source: Wu [3]. Reproduced with permission of John Wiley & Sons.

a foam can be the most critical part of a program. Regardless of the process, the development of foam involves the juxtaposition of gas generation and the development of tensile strength within the developing foam. The evolution of gas can be through the use of blowing agents or the *in situ* generation of CO_2 from the reaction of water with an isocyanate. Apart from the methods describe earlier, some success in pore size control it by the use of small amounts of stannous octoate.

The Special Case of Hydrophilic Polyurethane Foams

The following is based on our experience with hydrophilic prepolymers and the resultant foams. It should be noted however that these prepolymers are based on polyethylene glycol, TDI, and a little trimethylol propane. Any variation of that will make marginal differences in physical properties, but not so significant that it disqualifies the following discussion.

With regard to pore size and architecture, the conversion of the prepolymer to a foam by the use of surfactants is the most productive area of the research. We will discuss equipment to do this once your project has developed, but careful attention to hand-mixing is sufficient. The critical factor is in the production of a short-lived emulsion among the prepolymer and the water. The quality of the emulsion is controlled by the surfactant that is used. Table 2.11 is taken from the original product literature of Hypol 2000, but it still applies.

Physical and Chemical Testing

In the first section of this chapter, we reduced the chemistry of the first chapter into a functional device. Our ultimate goal is to craft those devices into something that will support cells or biomolecules. The device, however, must meet certain chemical and architectural requirements. We will discuss the chemical

Table 2.11 Control of cell size using surfactants.

Surfactant	Cell/foam type
Pluronic L-62	Fine cell/wicking
Pluronic L-520	Hydrophobic/medium cells
Brij 72 (ICI Americas)	Ultrafine cell/super soft
Pluronic L62 and P75	Large cells/“sea sponge”

Source: Adapted from Hypol Product Literature, Dow Chemical, Midland, MI, USA.

requirements in the last chapter. The architecture is described in the next chapter. In that regard, however, it is best that we focus on what methods and procedures are used to describe the material not qualitatively but with numbers. Accordingly, we will be explaining methods to define the material mechanically and structurally and, to some degree, its physical chemistry.

Inasmuch as the foam is a function of chemistry, we need to be sure the prepolymer is correct. A major contribution to the structure of the foam is the amount of NCO functionality. To follow is a description of the analytical procedure. Consult any from a number of texts to get the details.

As you develop your prepolymer, you will want to determine the reactivity. In this regard, it is the NCO content that is the preferred measure. The onset of foaming is an indirect, easy measurement but unreliable for research. The most reliable and convenient method to determine the total NCO (both end group and isocyanate NCO) is by titrations. A known amount of an amine (dibutylamine is used) is added to a weighed amount of prepolymer in solution. The amine reacts with the NCO groups, and the excess amine is titrated with a standard HCl solution to a bromothymol blue end point. The amount of NCO is calculated as either the equivalents per gram of prepolymer or the mass % NCO. The determination of the excess isocyanate is more problematic. The NCO groups are reacted with methanol, and the prepolymer is separated into its constituent parts in a size exclusion column. The methanol-capped isocyanates are the lowest molecular weight fraction. Consult any of several texts on this subject for a more complete description of the methods.

Analysis of a scaffold material is always part of the research. In conventional foams, it is the mechanical properties that are the focus. As a biological support structure, the suitability of a foam is more complex. This chapter is an attempt to position each of those factors into a comprehensive approach. The clinical success of a device depends on both. Engineers are already familiar with the structure requirements, and a set of standards and procedures have been established with quality systems requirements in mind. We will use those procedures, in some cases adjusted to the requirements of a medical device, as the foundation of a set of descriptors of a scaffold.

In this discussion we will devote much of the text on defining appropriate foams. It is consistent with our prejudice toward flow-through devices. Having said that, we know that foams can range from completely closed cell devices to full reticulation.

Our concern is, of course, the efficacy of the devices we are developing. It must be kept in mind however that we work in a field that strongly urges us to work with a quality system in mind. We must, even at the beginning of a project, recognize that we will be asked to confirm efficacy and safety aspects of our technology. While it is certainly more complicated in the medical device arena, the wise environmental researcher should keep in mind that whatever technology they develop will be closely examined economically and with safety as a significant concern. A wise researcher will consider these aspects from the beginning. While the concept of a device should not be constrained by these factors, once an idea is reduced to practice, quality and predictability are basic requirements. An integral part of that system is the design of the device or process, with suitable in-process and final product analysis.

There are certain things we know about scaffolds. While not exclusively, traditional foam structures are considered as viable. Within that category are open-cell foams because they allow for the spreading of cell cultures. Scaffolds intended for use in the repair of bone and cartilage are put under mechanical force, tissues intended for soft tissue require minimal tensile force, but compression needs to be addressed. Pore size, pore size distribution, interconnected cells, and so on are critical factors, as we all recognize. While not a part of all studies, the hydrophilicity should be part of a scaffold studied.

To that end and in the context of a scaffold for colonization or immobilization, we need to describe how one would evaluate the result of our chemistry and process. We begin this discussion with evaluating the physical properties of a scaffold.

The American Society for Testing of Materials (ASTM) has the largest inventory of approved methods of analysis and testing. Foams are covered in document D-3574 (Standard Test Methods for Flexible Cellular Materials—Slab, Bonded, and Molded Urethane Foams). The test procedures provide a standard method of obtaining data for research and development and quality control. The data obtained by these test methods are independent of the intended use. In this sense, the tests become an unbiased evaluation of the material with regard to the physical design, not the intended use. The interpretation of those results in the context of the success or failure of a scaffold is the first step in the development of a quality system.

While physical testing is important, the suitability of a scaffold for *in vivo* applications is equal, if not, more concerning. It would be counter-productive to a project if a scaffold design causes any one of several levels of incompatibility when implanted or comes into contact with blood. This is, of course, a continuous evaluation during the progress of a project. In that regard, there is a battery of tests that will be required before any clinical evaluation can be considered.

In this section we will give an overview of the testing requirements. Be aware however that much of the research that needs to be done is beyond the catalog of current testing.

Physical Testing

In the chapter to follow, we will discuss the physical requirements of a scaffold, whether it is for medical or environmental use. Engineers in the latter case are familiar with the testing as much as these have been standard procedures for decades. Equipment has been commercially available to test materials, and as we see in the ASTM manuals, rigorous procedures are established. Applying these procedures to the scaffolds for medical use is straightforward. These provide an accepted structure that we can use to evaluate our materials.

A unique factor however is that many of the foams we want to investigate have hydrophilic surfaces or in the extreme absorb significant amounts of fluid. Not mentioned in the standards is an appropriate pretreatment of the sample. You will have to use your best judgment to develop a pretreatment. Our work in hydrophilic polyurethanes requires a hydration step. The sample is soaked in warm water and dried in a towel just before testing.

We measure the % water in a fully hydrated sample and the swelling of a sample both linearly and volumetrically. This gives us the necessary size of a hydrated sample, indicating uniformity among a set of samples and, almost importantly for flow studies, any irregularities in the structure that would cause it to swell in one direction compared with another.

There are certainly other factors that might be valuable for your particular project, but for us this gives us a measure of the general category of hydrophilicity. However, one must use caution using this pretreatment for hydrophobic materials. Wetting of a surface is an imprecise characteristic and adding surfactants are counterproductive. For a more complete discussion of our pretreatment of a sample, refer to the earlier book. In that text we describe simple lab-scale tests that don't require expensive equipment [8].

Measuring the physical characteristics of the sample, as we said, is best reviewed in the various standards discussed earlier. Again, they are material specific and are not related in any way to how the materials are used. In our work, the relevance of a particular method needs to be judged individually. Tensile strength and elongations while important are rarely challenged in our medical application. As you will see in the next chapter, compression is usually more of an issue. Bone and cartilage applications require more strength that soft tissue.

We won't go through the procedures with the exception of this item. It has become a regular practice to describe interconnected pores in a qualitative manner. At best, researchers have shown micrographs of clearly open-cell foams as evidence without a quantitative measure. In the ASTM standard, however, it is a measure of this property. As an aside, reticulated foam manufacturers use this method to calculate pore size. The technique uses the amount of pressure it takes to pass a given volume of air through foam. Alternatively, the amount of air passing through a sample at a given pressure is determined. Figure 2.5 is a schematic of the technique.

The principle is obvious and the setup is not that difficult. We hope that this will replace the qualitative with the quantitative.

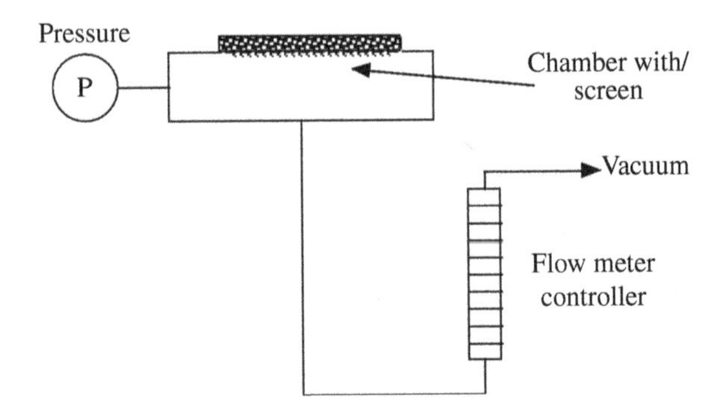

Figure 2.5 Apparatus for determining airflow through.

Biocompatibility Testing

In response to the rapid increase in materials considered for use as medical devices, a variety of cell culture methods are used to determine the suitability of a synthetic and semisynthetic material. As our focus is on the use of synthetic materials as scaffolds for cell propagations, it is clear that quantitative test procedures are needed to measure the nature the response of cells to the surface. While we would refer you to the following reference, it would be best that we delay this complex subject as we examine a number of studies conducted in the following chapters [9].

Process Equipment

For the most part the procedures and ingredients discussed earlier were designed to be hand-mixed. We have cautioned you that when architecture is a main feature of the device you intend, mixing past cream time destroys the natural (chaotic, if you will) structure of the foam. Even if it isn't destroyed, it invariably creates inconsistencies in the pores. Nevertheless, hand-mixing gives you a sensory appreciation of the process. As your project develops, however, you are going to need what is generally called a meter/mix device. The critical factors are a constant and precise metering of the ingredients and a near-instantaneous blending well before cream time. For reference there are two categories: high pressure and low pressure. The former is used for production level and was discussed in the first chapter. Low-pressure equipment is designed for prepolymers. Systems can be developed from benchtop to floor models, but they all have the same components.

Metering Pump

A metering pump is very important. They ensure a constant and continuous flow. Positive displacement pumps such as gear or piston pumps are preferred. Both types need to be fitted with a pressure relief system as "dead head" is messy and potentially dangerous (Figure 2.6). One must be aware that a positive displacement pump operates even when the lines are plugged. If the pump is strong enough, it will burst the tubing connecting it to the mixer. In our experience the following piping is effective.

Gear pumps are effective, and they supply a continuous stream of liquid without a pulse. We have used piston pumps with good success.

Mixing Head

There is a wide range of ways the prepolymer liquids can be mixed quickly. Among them are static, dynamic, and dynamic–static mixing. Of these a combination of dynamic and static mixing (a pin mixer) is preferred. A pin mixer has a rotor with short pins. It is inserted in a shell that also has pins. The liquids are pumped continuously into the chamber with the rotor is spinning. Machines can be made in which the ratios of components are fixed or variable. As research machines, variable would be a wise choice.

Tank/Material Retaining Container

A machine needs a container to feed the material to the mixer. The tanks need to have some pressure release mechanism as they are often pressurized. A nitrogen gas purge is advised for the isocyanate phase. You will need to choose tubes/pipes to transfer the liquids to the mix head.

Figure 2.6 Recycle to prevent dead heading.

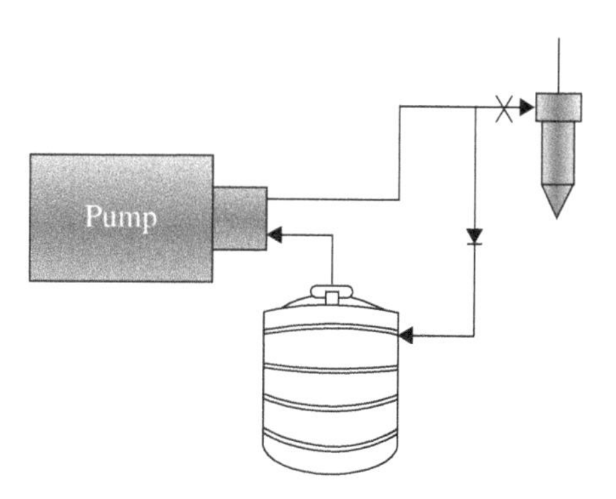

Heating capacity of the tank may be an option if you prefer to control the material temperature. You might consider agitators for heated tanks. For continuous pour processes, recirculation systems are important. Recirculation valve/lines help keep a constant temperature. It circulates the component material back to the heated retaining tank, so when your machine is not dispensing, use your equipment system builder for recommendations.

Machine Manufacturers

The following list consists of various manufacturers/suppliers selling meter–mix equipment. The recommendations are based on our experience. We have used and designed several of these machines. We have also built lab-based machines ourselves.

Ashby Cross http://www.ashbycross.com/
Sheepscot http://www.sheepscotmachine.com/
Graco http://www.graco.com/
Edge-Sweets http://www.edge-sweets.com/

References

1 *Hydrophilic Polyurethanes for Medical Devices: Science, Technology and Economics*, T. Thomson, CRC Press, Boca Raton, 1999.
2 *Polyurethanes Chemistry and Technology, Part II, Technology*, Saunders, J.H. and Frisch, K.C., pp 28–29. Interscience Publishers, John Wiley & Sons, Inc, New York, 1964.
3 The role of adsorbed fibrinogen in platelet adhesion topolyurethane surfaces, Wu, Y., Simonosky, F.I., Ratner, B.D., and Horbett, T.A., Wiley Periodicals 2005.
4 Synthesis of functionalized polyether urethanes and polyetherurethane ureas with linear and branched chains and their use in the modification of surface properties, Simonovsky, F.I., Porter, S.C., and Ratner, B.D., *Proc. Soc. Biomat.*, 20–46, 1997.
5 Confined compression experiments on bovine nucleus pulposus and annulus fibrosus, Perie, D., Korda, D., and Iatridis, J.C., *Journal of Biomechanics* 38, 2164–2171, 2005.
6 Structure property relations, Presented at the *12th Annual Seminar on Advances in Medical-grade Polyurethanes*, Technomics Publishing, Lynnfield, MA, Szycher, M., 1998.
7 *Polyurethanes Chemistry and Technology*, Part I, Saunder, J.H. and Frisch, K.C., pp 253, Interscience, New York, 1962.
8 *The ICI Polyurethane Book*, Wood, G., John Wiley & Sons, Inc, New York, 1987.

9 Biocompatibility of Polyurethanes, Madame Curie Bioscience Database [Internet], Marois, Y. and Guidoin, R. Landes Bioscience, Austin, 2000–2013.
10 New isocyanates from amino acids, Hettrich, W. and Becker, R. *Polymer*, 38 (10), 2437–2445, May 1997.
11 Biodegradable block copolymers as injectable drug-delivery systems, Jeong, B., Bae, Y.H., Lee, D.S., and Kim, S.W., *Nature* 388, 860–862, August 28, 1997.

3

Scaffolds

Fifty years hence ... we shall escape the absurdity of growing a whole chicken in order to eat the breast or wing, by growing these parts separately under a suitable medium.

—Winston Churchill, December 1931

Introduction

It may seem like a peculiar way to begin our exploration, but we think you will see that it is appropriate. The conversion of cells to tissue describes research goals of projects around the world. We have decades of experience in propagating mammalian cells, but we do not know how to incorporate them into what we refer to as tissue. Simply put, the development of a matrix within which mammalian cells can function is beyond our talents. It is this matrix that we refer to as scaffold.

In this chapter we will begin with an expanded definition of tissue to include other combinations using living cells supported by a structure. The definition goes beyond organic tissue but to include engineering structures. The reason for this will be clear. Engineering professionals, particularly those in environmental engineering and chemical processing, have used combinations of microorganisms and biomolecules supported by scaffolds as a matter of standard practice. We will present the case that we know a lot about this subject and the properties of engineering scaffolds. The challenge that we face is to apply those principles to medical devices.

This chapter deals with scaffolds and the engineers have a lot to teach us. In the next chapter, those roles will be reversed. The medical community and to some degree engineers have become very skilled in the topic of

Polyurethane Immobilization of Cells and Biomolecules: Medical and Environmental Applications, First Edition. T. Thomson.

immobilization. Unlike bacterial cells that have a genetic ability to stick to surfaces, most mammalian cells develop intercellular adhesion (and even communication), but are not "designed" to stick to most artificial surfaces. As we will discuss, human cells develop an extracellular matrix (ECM) for cell adhesion to themselves and surfaces. It is this method that is part of the conversion of cells to tissue. We will look at ECM in some detail as the model for scaffold development.

Remembering that this book has two readers, medical researchers and environmental engineers, let me propose a unified definition among the two areas of research. The definition attempts to get researchers to think of environmental remediation systems that use microorganisms attached to a scaffold as hybrid tissues. We want medical researchers to see their cell/scaffold constructions as engineered structures. For the purposes of this text, we are defining a bioscaffold that has two primary attributes:

1) It is an open architecture that allows for the passage of a fluid through it with high surface area and a minimal resistance to flow.
2) The internal surface is treated with a biologically active molecule or colonized by living cells such that a fluid passing through it is chemically changed.

Defined this way it doesn't matter if the fluid is municipal waste or blood. This is the point we want to use to begin the discussion of scaffolds. Environmental engineers have had to juxtapose the two requirements from functional and operational points of view for a 100 years or so. Open architectures and high surface areas relate directly to cost and efficiency. The goal is to pass as much fluid through a scaffold at high flow rates and minimal internal pressure as possible. This is summarized in part into properties referred to as empty bed residence time (EBRT) and pressure drop. EBRT relates to the reaction rate, and the pressure drop defines the size of the pump needed for the required flow rate. The active surface defines the chemistry. The development of a functional surface has many implications, all under the general subject of immobilization, the subject of the next chapter.

Using this broader definition, not limited to medical applications, allows us to review how engineers have used the cell/scaffold relationship to solve environmental problems. This approach allows us to juxtapose hybrid artificial organs and the treatment of agricultural pesticide runoff as engineering problems. We refer to this construction as a bioscaffold. The surfaces of a bioscaffold can be colonized by animals, plants, yeasts, molds, or algae. In addition, the internal surfaces can be activated by enzymes, antigens, antibodies, and proteins.

With that as an introduction, it is appropriate that we begin our discussion of the scaffolds used as the structural component of a bioscaffold with the engineers and researchers who are most involved in the subject. While there are many configurations used to treat waste, most appropriate to our discussion is the technique referred to as biofilters.

Before we get to the serious stuff of scaffold development, it is useful to report on a curiosity in this field. A group at Maastricht University in the Netherlands "successfully" produced a synthetic hamburger from cow muscle cells. As reported in *The Guardian* (August 5, 2013), Dr. Mark Post and his team used these cells to grow 20 000 muscle fibers in a gel-like growth medium. The fibers were then pressed together, colored with beetroot juice, and mixed with saffron, breadcrumbs, and some binding ingredients to form the burger. Even in this simple example, which clearly cannot be considered an organ, a scaffold-like material ("some binding ingredients") was necessary to convert the cells into tissue. Clearly, although the example was tissue, it does not qualify as progress in the development of an artificial organ. In order to make progress in this area, one must add the properties of permeability and biocompatibility.

Bioscaffolds

Bioscaffolds for environmental remediation (biofilters) are defined as a structure incorporating living cells, typically bacterial, through which a fluid passed. The purpose of course is to remove contaminants or otherwise change the chemical nature of the fluid. The primary focus will be with systems that immobilize the organism. By immobilize we mean that the organism is affixed to a scaffold by an electrostatic attraction, encapsulation, or a covalent bond. The discussion is separated into two segments. In order for a system to be called tissue, it must, in the minimum, be a colony of cells dispersed in or on a support structure. As we said it must also be permeable and it must in some way change the chemical nature of the fluid to which it comes into contact. As an aside, you will notice a prejudice for flow through applications. There are other constructions that depend on bacteria to affect a chemical change. Stirred tank vessels are common to municipal waste treatment. These are typically colonized by free-flowing bacteria. While some might consider the tanks as scaffold, this would remove it from medical device consideration. Again, biofilters use microorganisms affixed to a porous medium to break down components present in a fluid stream passing through it. The porous medium is typically relatively inert, and its primary purpose is to give the organisms a large surface area to colonize. The effectiveness of a biofilter is governed by the properties of the support medium. This includes porosity and the ability to host an appropriate microorganism without interfering with its function. Critical biofilter operational parameters include flow rates and residence time.

There are two types of biofilters. The classic biofilter is intended for gas (air) treatment, while the so-called biotrickling filters are used for liquids, typically water. In its use in environmental remediation, both gas and water pass through the filter. This ensures that the organisms are supplied with both waterborne nutrients and oxygen. The goal of most biofilters is to create biomass. It is important to note that the purpose of the human liver, in part, is to metabolize contaminants, while a biofilter typically captures them in the form of biomass. Thus, while the definition, in our opinion, holds, it goes without saying that *in vivo* processes are more complex. We do, however, suggest that biofilters give us guidance as to bioscaffold design.

To that end let us examine how a biofilter is designed. The first step in that process is a definition of what the system is intended to treat. We will go into a number of treatment strategies, some from the literature and our research, but for the purpose of clarity, let us assume that we will be defining a system to treat municipal waste. There are pretreatments such as settling and screening, but our focus is on the treatment of the liquid fraction of the waste stream. The goal is to design a biofilter that can remove the organic matter. This is, of course, the combination of a scaffold colonized by a system of microorganisms and an architecture that efficiently contacts the waste with the activated surface.

To define the system, the engineer begins with certain critical parameters. The volume of liquid per unit time determines the size of the operations. The goal of the process, in terms of the quality of the effluent, is usually set by local governments. If it is to be drinking water, that is one thing. If it is to be pumped into a nearby waterway, that is another. The nature of the water to be treated needs to be defined, as is the source of the biological material that will metabolize the waste. Capital and operating costs need to be estimated. This list continues to include the footprint of the site, local utility services, and local community considerations. (As an aside, the last three factors apply to an artificial liver as well.) To say the least, the issues range from technical to political. We, of course, are only concerned with the engineering design. As we said we will review a number of illustrative operations including the work that we sponsored at the University of California. Before we begin with examples, however, it is useful to talk about relevant terminologies. In our opinion, the most relevant is called the EBRT. In as much as chemical and biological processes are not instantaneous, the contact time of the surface and the pollutant is finite. Thus one must be aware that the flow rate through the biofilter must allow for the reaction to take place. In as much as the absolute volume of the scaffold on which the microorganisms reside is inactive, that must be included in the calculation of the contact time. The calculation therefore is to subtract the volume occupied by the scaffold from the total volume and divide that by the flow rate:

$$\text{EBRT} = \frac{\text{Volume of the vessel} - \text{volume of the scaffold}}{\text{Flow rate}}$$

Consider a vessel of one cubic meter (m^3) packed with commercial reticulated polyurethane (PU) foam. The foam has a void volume of about 97%, and so the absolute volume is 0.03 m^3, leaving an empty bed volume of 0.97 m^3. At a flow rate of 1 m^3/min, the residence time would be 0.97 min.

Now consider the same vessel packed with 1 cm diameter stone, a common biofilter packing. The void volume is approximately 50%, and therefore the empty bed volume is 0.5 m^3. At the same flow rate, the residence time is now 0.5 min. If the reaction rate requires more than 0.5 min, the device would not be sufficient.

You will have probably noticed that this simple calculation ignored the effect of surface area. The scaffold material is colonized on the surface only. It doesn't take much imagination to assume the stone in the aforementioned example has a lower surface area than the reticulated foam. Thus while the residence time is lower, the bacteria-coated surface is also lower, thus decreasing the overall efficiency of the vessel beyond the decreased EBRT.

There are other factors that need to be addressed as you might imagine. We will cover many of these in the paragraphs that follow. We wanted to highlight the ones that are critical to medical devices as well. Among these is the concept of pressure drop. It has several names (mass transport is an important one), but the concept and its importance is simple. If one is to pump a fluid through a bioscaffold at a given flow rate, the pressure applied to the inlet of the filter needs to be controlled. The design of the filter material itself needs to account for the surface area and void volume. These two factors in part define the kinetics. External factors like the amount of fluid that needs to be treated per unit time are then applied to determine the optimum flow rate. The pressure needed at the inlet is then established, assuming nothing changes. In an engineering environment, that establishes the size and power of the pump. *In vivo*, however, the pump size is fixed, and so the filter must be modified to increase the EBRT, for instance. For the engineer, it is the pump that is typically sized, but within some constraints. Larger pumps are more expensive and the operating costs are higher. This needs to be included in the cost of the project.

With regard to the first two requirements, Muir [1] reported the surface area and estimated void volume of a number of common packing materials used for biofilters. Some of these are trade names (Table 3.1).

While the table is clear with regard to the void volume and surface area of the common packing materials, PU foams, the focus of our research, are more complicated, therefore offering more opportunities. To illustrate, the so-called open-cell foams have high surface areas and void volume. Both would be at the top of any list in both categories. However, as we have discussed, the pressure drop depends on the degree to which the cells are interconnected. Still further, without taking steps to adjust to this, the foams are compressible and therefore tend to collapse under pressure. This problem is fairly easy to correct but nevertheless needs to be considered.

Reticulated foams were developed in part to mitigate this problem (Figure 3.1). Nearly all the windows between the cells are removed by a

Table 3.1 Surface area of common scaffold materials.

Material	Surface area (m^2/m^3)	Void volume (%)
0.5″ rock	420	50
1″ rock	210	50
2″ rock	105	50
0.5″ ceramic	364	63
0.5″ carbon	374	74
0.5″ Berl	466	63
1″ PVC saddles	249	69
1″ PVC pall rings	217	93
1″ Raschig rings	190	73
Surfpac	187	94
Flocor RC	330	95
Cloisonne PVC	220	94

Source: Adapted from Szycher [2].

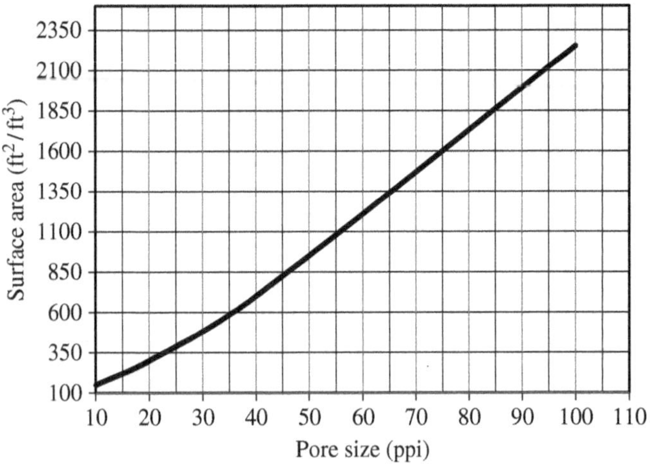

Figure 3.1 Surface area by pore size of a reticulated foam. *Source:* Adapted from FXI Corp. [3].

post-foaming process. This is discussed in Chapter 1. The foams were developed specifically as air filters. As such both pore size and uniformity are critical, as was pressure drop. Figure 3.1 is taken from product literature from a major reticulated foam manufacturer, FXI Corporation of Medina, PA, USA.

Table 3.2 Range of pore sizes within a pore size grade.

Pore size grade (ppi)	Minimum	Maximum
100	80	110
80	70	90
60	51	65
45	40	50
30	25	35
25	20	30
20	15	25
10	8	15
3	3	5

Source: Adapted from FXI Corp. [3].

Compare this with the surface areas of the more common materials used for biofilters. Void volume is also very high and, surprisingly, is not a function of pore size. This seems counterintuitive, but the void volume is constant around 97% regardless of the pore size. This is understandable, however, when you see that the density of the foam is also constant across the pore size range.

One other factor is also important to know, the uniformity of the pore size. Table 3.2 shows the pore size of a reticular foam approaches monodispersed. The data is also from FXI Corp.

In general, the treatment of municipal waste requires high flow rates but the residence time in a biofilter cannot be sacrificed. The designer uses the pump power and flow rate as well as the size of the equipment to control residence time in the filter. Many of the natural and synthetic materials in Table 3.1, when packed in a column, become more densely packed or, due to mechanical action, break and change the residence time or resistance to flow over time. The buildup of biomass also contributes to process control problems. In an example of our research, we built a biofilter for the removal of toluene and another for the removal of hydrogen sulfide. The former builds biomass that caused the packing to collapse. The degradation of hydrogen sulfide however does not produce biomass and the biofilter was successful. Both biofilters were packed with a composite reticulated foam. We will use this project as an example.

Examples of Biofilter

We think it would be instructive to look at a few examples of biofilters in commercial operation. We will also explain some of our research on the use of a composite reticulated foam. Engineers will see this as simple descriptions of systems to remove pollutants from industrial or municipal operations. We would ask the medical research community to think of these as very large tissues. We will describe the flow of a fluid through a biologically active scaffold to accomplish a chemical or constituent change. This is not functionally different than the liver or kidney. The liver breaks down harmful or toxic substances and then excretes them into the blood. By products excreted are filtered out by the kidneys and then leave the body through urine. The liver also metabolizes drugs, making them inactive or easier to excrete from the body. This last function is an important consideration in new pharmaceutical research. Drugs injected into the blood immediately pass through the liver before they enter the region for which they are designed. This is called "first path." The stomach has a similar function but more or less analogous to a stirred tank reactor, that is, not a biofilter. There are enteric delivery systems that protect the drug from the stomach environment but are acted upon by the gut. The drug then enters the lymphatic system, and therefore it is not "first pass." The gut in this sense acts as a tube reactor with high surface area colonized by bacteria ("normal flora").

From this explanation we think you can see the argument that biofilters can be functionally considered a tissue. In any case, we think it would put the medical community in a useful frame of mind, while we focus on large industrial processes.

The liver has productive functions as well. It manufactures cholesterol, which is used to make bile, a fluid that aids in digestion. Cholesterol is also needed to make certain hormones, including estrogen, testosterone, and the adrenal hormones. The liver also manufactures proteins needed by the body for its functions. For example, clotting factors are proteins needed to stop bleeding. Albumin is a protein needed to maintain fluid pressure in the bloodstream (flow rate vs. pressure drop). Sugars are stored in the liver as glycogen and then broken down and released into the bloodstream as glucose when needed.

Like the liver, industrial biofilters by their nature don't eliminate, but rather metabolize. While for the most part they are considered detoxifying, in another sense they make product. The product may be considered fertilizer but nevertheless it is a product. This again supports the concept that biofilters are tissue. This is easier when you think of a robot that converts sugar to alcohol that is burned to make electricity to operate the gears and actuators.

With that, we will look in detail at examples. The design of an industrial biofilter for remediation begins with the specification of the proposed process. Not to belabor the point, we will examine the process to define the biofilter and juxtapose it with an examination of a malfunctioning liver.

Elimination of Tobacco Odor from a Cigarette-Manufacturing Plant

A tobacco processing plant was required to reduce odors by 90% or to a level less than 100 odor units [4]. A pilot plant study confirmed this was possible within the space available on the roof of the plant. The nature of the emissions was such that there was a concern that the proposed system could become clogged with biomass. The pilot plant determined that this was not likely. The system provided for intermittent watering of the scaffold.

The scaffold used for the system was a reticulated foam cut into 40 mm cubes. The pore size was about $10\,m^2/m^3$ and the density about $20\,kg/m^3$. The surface area was relatively low for a reticulated foam. This was selected because the air would flow freely through the bed, but not necessarily through the foam. Therefore very large pores were specified to mitigate this. By comparison, if the air was forced to go through the foam, smaller pores and high surface areas could be used. This is explored later when we discuss our work with the University of California. For practical reasons however, large pores and cutting the foam into cubes were required for this project. Watering of the scaffold was done at 5–15-min intervals.

Acclimation took place over a period of 2 months at which time odor removal was greater than 90%. In an examination of one of our projects, the scaffold was colonized with local municipal waste, making it immediately active.

Table 3.3 describes the system.

Table 3.3 Tobacco plant biofilter.

Scaffold medium	Reticulated polyurethane foam
Amount of scaffold	$500\,m^3$
Airflow	$160\,000\,m^3/h$
Empty bed residence time	11 s
Pressure drop	400 Pa
Average bed temperature	40°C
Input odor level	800–5500 OU
Controls	Temperature, pressure drop, and water level
Installation cost	$3 mm
Operating cost	$0.07 per $1000\,m^3$
Performance	>90% removal

Source: Deshusses *et al.* [4]. Reproduced with permission of Taylor & Francis.

Treatment of VOCs from an Industrial Plant

Various emissions from a wastewater plant contained low concentrations of volatile organic compounds (VOCs) [5]. Among them were toluene, xylene, and chlorinated hydrocarbons. A biofilter was designed based on layers of a combination of a 50:50 mixture of compost and polystyrene spheres, followed by limestone to neutralize acidic compounds. Clogging by biomass was a significant problem. Pressure drop was controlled by replacing the top layer of the scaffold periodically. The degradation of the chlorinated hydrocarbon presented a pH problem (levels of pH 2 were experienced), but this did not seem to affect the function of the filter.

Table 3.4 described the system.

The Liver as Biofilter

The liver plays a critical role in numerous functions in the human body. This includes regulation, decomposition, synthesis, production, and detoxification. It produces bile, which aids in digestion. Liver cells (hepatocytes) regulate a wide variety of biochemical reactions, including the synthesis and breakdown of small and complex molecules (Table 3.5).

Thus, we have described a couple of traditional biofilters and then, hopefully we compared those successfully with the human liver to further emphasize the notion that a biofilter can be considered an organ and vice versa, considering this closing comparison (Figure 3.2).

Table 3.4 Hydrocarbon removal.

Air stream	Exhaust from wastewater treatment plant
Packing	50:50 compost and polystyrene beads
Airflow	60–75 000 m^3/h
Empty bed residence time	2 s
Pressure drop	2 kPa
Bed temperature	30°C
Pollutants	Aromatic, aliphatic, and chlorinated hydrocarbons
Controls	Flow, temperature, pressure drop, efficiency, moisture
Investment cost	$4 mm
Operating cost	$1.5/1 000 m^3
Performance	80% removal

Source: Deshusses *et al.* [4]. Reproduced with permission of Taylor & Francis.

Table 3.5 The liver as biofilter.

Fluid stream	Blood
Packing	Hepatic cells
Flow	1.7 l/min [6]
Empty bed residence time	60 s (estimated)
Pressure drop	0.8 kPa
Bed temperature	38°C
Pollutants	800 separate functions
Controls	Flow, temperature, pressure drop, efficiency, moisture
Investment cost	0
Operating cost	0
Performance	Fully functional

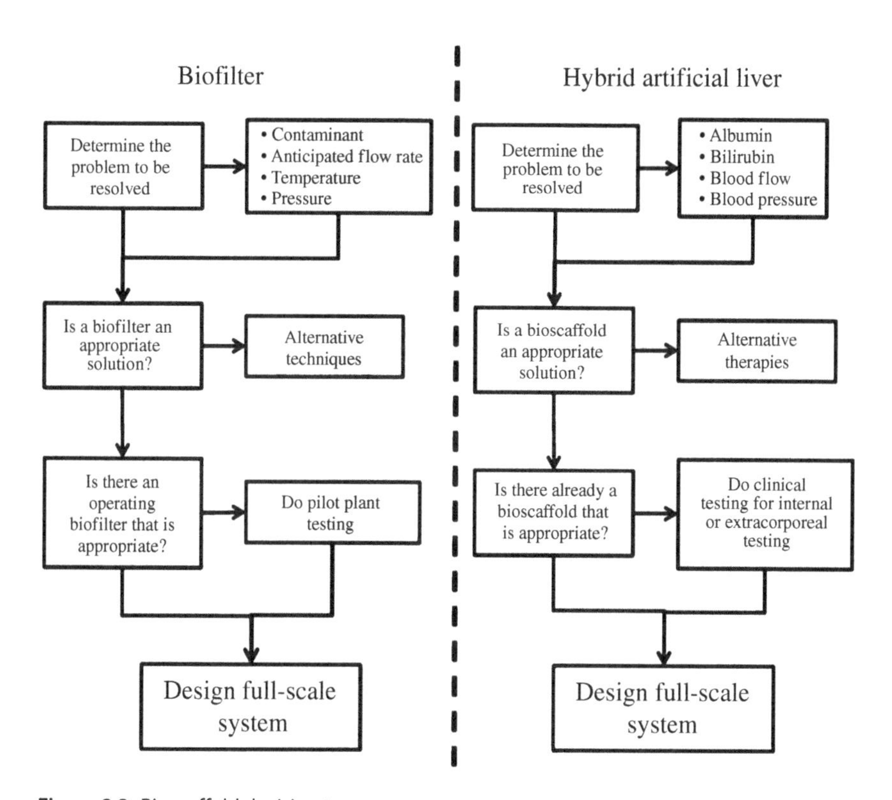

Figure 3.2 Bioscaffold decision tree.

With that said, we are now ready to discuss the specifics of the scaffolds for remediation and medical devices. We begin with the former.

Scaffolds for Medical Applications (*In Vivo* and Extracorporeal)

In spite of the great variety in the materials and methods employed in the fabrication of tissue engineering scaffolds, the principle for the scaffold designing remains clear. The scaffold should be designed by mimicking the native ECM as much as possible, both in terms of chemical composition and physical structure. In nature however, the ECM is produced by the cells themselves, while the task of the tissue engineer is to develop a scaffold and then immobilize cells or biomolecules. Nevertheless, the function of the ECM is far more complex than just providing a physical support for the cells. It provides a substrate (scaffold) containing proteins for cell adhesion and regulates cellular growth and functions by presenting different kinds of growth factors to the cells. An ideal ECM-mimic tissue engineering scaffold should duplicate as close as possible to the chemical function. In the case of the immobilization of biomolecules (enzymes, antibodies, antigens, etc.), the following is a short list of requirements of a scaffold:

- The linkage should be strong enough, preferably covalent, to avoid displacement by biological media.
- The binding site of bioactive molecules must not deactivate.
- The bioactive molecules should not denature or change configuration upon binding.
- The anchoring reaction should only react with specific groups on the bioactive molecules.

Thus it is the scaffold that, in part, defines the tissue, but it is the nature of the scaffold that gives it function. Each researcher will have his or her working definition of an ideal scaffold. Whether the purpose requires vascularization has a profound effect on the properties. We were involved in a project to develop a PU hydrogel for the repair of certain spinal injuries. Blood contact was not an issue and this made the problem less complex. If, however, the goal was an artificial liver, blood contact and permeability would be first on the list of required properties.

Using the most difficult case of producing an artificial organ (liver, pancreas, etc.), we will describe what we and others consider the properties of the ideal scaffold. It is the purpose of this book to explain why we feel that PUs and derivatives thereof may be a useful candidate.

The Liver Model

Periodically we will refer to the liver model. While applications for hybrid organs and tissue in general are broad, the human liver offers a complex model with a strong demand. Liver transplant survival rates have improved such that 5-year survival rate of 75% is typical. Transplant surgery is a durable therapy for all forms of end-stage liver disease and for some malignant conditions. The success of such treatment has resulted in a progressively increasing demand. However, at the same time the availability of donor organs has diminished, resulting in the number of potential recipients for liver transplantation exceeding organ supply. Several strategies have been explored to increase access to liver transplantation, including whole and splitting livers as well as liberalizing the donor criteria.

Concurrent to surgical methods, research around the world has focused on hybrid liver constructs that either supplement the function of a natural liver or even replace the organ. Culturing of natural hepatic cells is sufficiently developed to direct the effort toward the development of a support structure (scaffold) on and within which the cells can be cultured. Referring back to the beginning of this chapter, the goal of these studies is to make liver tissue. As such it must fulfill certain physical and what we call architectural considerations.

It is important that the reader recognize that we are chemists. As such, our experience extends across many application markets. In this book we focus on environmental and medical devices. We have tried to show that they are connected by a need for a colonized scaffold. We, however, have to admit, while we hope to catalyze innovation in both fields of study, we make no claims as to the ultimate developments. Especially in the medical arena, the issues are so complex that we can only suggest a path, but have no ability to move along that path under our own power. The liver is a case in point. When we speak of a liver model, we are envisioning an extracorporeal device that is shunted into the femoral artery to ***assist*** a damaged liver. In this context we avoid the physical complications of implantable devices, vascularization, and biodegradability. The liver model that we do envision is a high surface area, biocompatible, low pressure-drop scaffold colonized by immobilized autologous hepatic cells. Biodegradability is a solvable problem. PU chemistry has multiple ways to degrade. From the urea linkages to polyols made from natural products (e.g., castor oil), the chemist only needs some medical researchers to at least define the chemistry. The architecture is problematic, but we feel that information in this book will help define an appropriate target. Hitting the target with a biodegradable PU would probably be the simplest goal in the development of an implantable organ. On a personal note, we think it was a mistake to include the goal of implantability in scaffold development. Biodegradability is comparatively easy. Building a functioning scaffold is difficult. So to repeat, our liver model is a biodurable device.

The Extracellular Matrix as Scaffold

As product designers, we naturally begin a project of this type with the physical requirements. Scaffolds in general are open structures of significant strength. Surface area can be an important consideration. If the requirements stopped there, however, a reticulated PU foam might be the clear choice of materials. It has all of those characteristics. As we have learned, however, a scaffold for our liver model is much more complicated. Along with the structural properties, the scaffold must fulfill certain chemical requirements. Along with the concept that it must "do no harm," it must operate within a synergistic environment. That is, it must contribute to the development of tissue, not just allow it to develop. Recognizing this it is appropriate to begin with a discussion of the natural process. While the effect is in part immobilization, the subject of the next chapter, the scaffold function, is dominant enough to include it in this discussion.

Cells in the body are held in close contact with one another and develop intercell communications. The assemblage is what we have called tissue. The medium that facilitates this construction is referred to as the ECM. It is a secretion from the cells of proteins and carbohydrates. The ECM is unique to each tissue and organ. For example, in tendons a fibrous ECM makes the tissue resistant to stretch, while in cartilage and bone, the matrix resists compression. Also in the bone the ECM includes a mechanism for mineralization.

It not only plays a mechanical role but also affects cell function. It is the natural scaffold material for cell attachment, proliferation, and differentiation. This is particularly important in early growth.

Components of the ECM, such as collagen, laminin, and fibronectin, have been used as surface coatings for biomaterials to facilitate cell attachment and growth and to promote biocompatibility. As we said, there are as many types of ECM as there are types of tissue, each having a specific function. The basement membrane, for instance, is of particular interest to our topic. It is a type of ECM that is between the cells and surface to which it is attached. It defines the conformation of the cells in relation to the supported surface.

The influence of the ECM to cell proliferation and spreading is strongly influenced by the surface. We will discuss a work done in Japan on a reticulated foam-based hybrid artificial liver [7]. Micrographs of the foam containing the liver cell show aggregation of the hepatic cell into a spheroid. It is clear from the micrographs that while the cells propagated, they did not spread. In fact there was little association with the scaffold. While we feel this research is groundbreaking, consideration needs to be made to the reticulated surface to, in part, mimic a basement membrane derived from a natural hepatic organ. The basement membrane is a critical factor in the choice of an ideal bioscaffold for tissue engineering applications. Additionally, in as much as ECM is specific to the cell type, understanding of the relationship is important.

This specificity of the ECM to the cells was investigated at the University of Pennsylvania. In the study ECM scaffolds made from the urinary bladder, small intestine, and liver of an adult pig were examined to determine the effect of the basement membrane complex on the growth of human endothelial cells and fibroblasts [8]. The surfaces were cleaned, leaving just the scaffold material. Staining of the scaffold showed differences in the composition of the surfaces including the type of collagen. Colonization revealed that epithelial and fibroblasts infiltrated the tissues differently.

Using our liver model as a goal, we are aware that the construction of the device is more than a physical scaffold. It must also participate in the proliferation and health of the hepatocytes. We also recognize that the two aspects are not sharply differentiated. For the purposes of clarity, however, we will separate the issues into, first, this chapter on the physical scaffold and, second, in the next on how the scaffold acts chemically. Nevertheless, we recognize that the success of the project depends on the combination of the two. With that let us direct your attention to research on the design of a physical scaffold. We will give examples of our research and that of researchers around the world.

The Physical Scaffold

Requirements for a scaffold include factors ranging from surface characteristic to architecture. We will rely on the literature for this discussion, but we will use our prerogative to examine what a scaffold for our liver model might look like. In that regard, the device will require not only compatibility with the hepatic cells, of course, but also compatibility with the fluid flowing through it, that is, blood. The scaffold must be hemo-compatible or at least compatible with plasma. Also related to compatibility is the association of the cells to the scaffold surface. Among these are provisions for anchorage. We talk about this more completely in the next chapter, but while there must be some level of attraction of the hepatic cells to the surface, it cannot be so strong as to affect the natural conformation of cell-to-cell association.

We have mentioned on many occasions that architecture is as important as chemistry. Our purpose here is to describe a scaffold to meet the physical requirements for a flow-through device. Such a device will nourish the cells as they proliferate and serve as a bioactive surface to metabolize fluids. It is critically important to recognize however that our goal is not to create tissue, but to build an appropriate support structure for the tissue to develop. In that regard, flow, interconnectedness, and void volume work in combination to support the purpose. Flow is dependent on the independent variables of void volume and interconnectedness. Not one or the other, but both.

Void volume has an additional consideration. During the development of the cell colony, there must be sufficient volume not only for the proliferation but also for the conformation of the expanding cell population. We must remember that while we must mimic ECM, we are not making an ECM, but a scaffold on which cells can develop a natural matrix that meets its cellular purpose. All our artificial scaffold can do is provide a starting point. Thus maximizing void volume is critical.

In summary, we are not making tissue, but are creating an environment for natural tissue to develop. In this sense, it would appear to be appropriate to build a scaffold that would disappear when it has served its purpose. While conceivable, timing, degradation products, and other factors have slowed progress to that overall goal. Thus the design of a scaffold for our liver model is a biodurable extracorporeal device, leaving implantable devices for the next generation of researcher. Dialysis equipment continues to save lives, while an implantable kidney is developed.

Despite this there is a wealth of research on what an ideal scaffold might look like. We will review what we feel are the most significant opinions on this subject and conclude the chapter with examples.

While we will prejudice the discussion to medical science, the engineers will notice that what we talked about previously is a distillation of how an engineer will describe scaffold. Many of the medical applications are cell-type specific, but the scaffold design works as well across the fields of study.

Design of an Ideal Scaffold

Implantable versus extracorporeal scaffolds differentiate the discussion. Our liver model emphasizes this point. It affects the choice and option of materials among other things. The goals of each, on the other hand, have commonalities that make it appropriate to begin with a generalized point of view.

The overall aim is the development of biological substitutes that restore, maintain, or improve tissue function. This is true whether it is implantable or not. We know now that the engineering approach to control tissue formation is in three dimensions. It is a highly porous scaffold with a high degree of interconnectedness. The scaffold provides an environment that in some aspects serves as an ECM even if temporarily. Specifically, it should support cell attachment, proliferation, and perhaps differentiation. Various materials have been exploited as scaffolds including metals, which have excellent mechanical properties but present a challenge in architecture. Ceramic materials such as hydroxyapatite (HA) or calcium phosphates are being studied for mineralized tissue engineering, but again porosity is problematic. We will discuss research on both of these materials. Polymers would appear to have the highest potential and PUs have assumed an important position. While extensive research has gone into adapting the most common polymer systems (polyethylene, polypropylene, polyvinyl chloride, nylon, etc.), we feel PUs have a critical role to play. Unlike these other polymer systems, simple changes in chemistry, available with undergraduate knowledge of chemistry, change the material from elastomers to foams and foams from closed cell to open cells to fully reticulated foams. Still further, as we will discuss in the next chapter, immobilizing biomolecules is as simple as mixing an aqueous solution with a PU prepolymer. We will discuss an exciting development in which a polypeptide is inserted into the PU molecule itself.

Again the role of a synthetic ECM for the propagation of cells requires that the material emulate certain features of the natural ECM. It is, however, unnecessary and even impossible for a scaffold to entirely duplicate the ECM. Referring to our liver model, even if natural healing is possible, it is probably too slow. In this case an accelerated process might be lifesaving. Therefore this aspect must be part of the design of a scaffold. Porosity, pore size, interconnectedness, and so on are necessary for accelerated tissue repair.

We conclude this section with an aspect of scaffold design that is not well reported. As we develop a generalized scaffold, we will focus on scaffold design for each tissue type. We have discussed the specialized nature of ECM on cell type. In some cases, the elastic properties of the scaffold have been shown to be an important characteristic. Engler *et al.* [9] showed that stem cells were sensitive to elasticity. Soft matrices encourage nervous system development, stiffer matrices assist in muscle development, and rigid matrices are for bone growth. Cells commit to the lineage specified by matrix elasticity. According to the authors, "inhibition of non-muscle myosin II blocks all elasticity-directed lineage specification—without strongly perturbing many other aspects of cell function and shape." The results of the study appear to have significant implications for understanding physical effects of the *in vivo* environment.

Yang *et al.* summarized the search for an appropriate scaffold as based on the belief that hepatic cells can be taken from a patient and then be cultured in a carrier (scaffold). The resulting tissue construct is then implanted back into the patient. To practice this technique, a highly porous artificial ECM or scaffold is needed to accommodate cells in three dimensions. There is another application, however, that deserves our attention.

Drug Discovery

We have not mentioned this before, but there is another reason to focus on extracorporeal devices. The concept of three-dimensional geometries raises this issue. While the ultimate goal is to serve as a prosthetic device for patients with liver insufficiency, there is an intermediate goal to develop a devise that would function as a test cell for exploratory drug testing. We will spend some time on this subject as it serves as a model for a clinical-level extracorporeal device. Nearly important is the unmet need for a device to determine the toxic effects of new drugs. That is to say that if we develop a flow-through test cell containing functioning hepatic cells for toxicity testing, we have simultaneously developed an extracorporeal hybrid artificial liver. Again we feel this should be the first and second steps in the development of an implantable device.

The needs are clear. The cost of drug development has risen markedly in the past 30 years, with studies now reporting values exceeding $1 billion for each new drug. Pharmaceutical and biotech companies spend about $60 billion on research [10]. Liver toxicity *during clinical trials* accounts for over 40% of the withdrawals. One company estimated that withdrawals during clinical trials will increase to $2 billion. These and other factors make the development of new drugs unaffordable for both developing companies and consumers. In addition, liver toxicity accounts for about 27% of drug withdrawn from the market in the period from 1960 to 2002. Lastly, drug-induced hepatic injury is the most frequent reason for the withdrawal from the market of an approved drug, and it also accounts for more than 50% of the cases of acute liver failure in the United States today. More than 75% of cases of drug reactions result in liver transplantation or death [11].

The unmet need is clear. The current testing is all but ineffective. It is also clear that a logical solution is to test new drugs on human hepatic cells. The availability of accurate informative *in vitro* assays is an increasingly important challenge facing the pharmaceutical industry. Poor predictive value of existing *in vitro* tests places great emphasis on the development of more realistic cell culture models.

Currently, cell-based *in vitro* assays are a key component of drug discovery research. However, traditional cell culture environments are far removed from real-life tissues. Cells grown in two-dimensional monolayer culture cannot develop natural morphologies or communicate efficiently cell to cell. They lack many of the physical and chemical cues that allow them to function *in vivo*. It is the growth substrate that affects cell function. Thus the development of a three-dimensional culture system for drug development is an important part of this process. Excellent reviews of the subject are by Barcellos-Hoff *et al.* [12] and Schmeichel *et al.* [13].

It is useful we think to consider both preclinical drug testing and the development of an extracorporeal artificial organ and the same project (except perhaps in physical size). Extracorporeal devices had the option of autologous cells. The development of an appropriate device itself has two aspects, the materials and the architecture. We have discussed this briefly, but in the following we elaborate.

We will now return to Yang *et al.* for a discussion of the materials that are considered appropriate for scaffolds and supplement those data with examples.

Materials of Construction

The Consensus Conference of the European Society for Biomaterials is defined a biomaterial as a material intended to interface with biological systems to evaluate, treat, augment, or replace any tissue, organ, or function of the body. Typically, three individual groups of biomaterials, ceramics, synthetic polymers, and natural polymers, are used in the fabrication of scaffolds for tissue engineering. Each of these individual biomaterial groups has specific advantages and, needless to say, disadvantages, so the use of composite scaffolds comprised of different phases is becoming increasingly common. In this discussion it is important to differentiate relatively hard tissues, bone, and cartilage with softer tissue. As we learned from Engler *et al.* [9], the elasticity is to some degree a predictor of stem cell fate. For this reason we will separate the discussion on, for lack of a better word, stiffness. In addition, there is the issue of blood contact that differentiates our discussion. That is not to say that we will not ignore issues like surface area and interconnectedness.

Ceramics

Ceramics are commonly used in tissue replacements. Typical ceramics are alumina, zirconia, calcium phosphate including HA, and bioglass. Hip and knee replacement surgery is based on these materials and has remarkably improved the quality of life of many people. However, metals and ceramics have two major disadvantages when applied to some tissue engineering applications. Metals and many ceramics are not considered biodegradable. Exceptions as we will show are the so-called bioceramics in various forms of tricalcium phosphates (TCPs). Converting these into a suitable scaffold is problematic but, as the following examples show, is workable. Although not generally used for soft tissue applications, there has been widespread use of ceramic scaffolds, such as HA and TCP, for bone regeneration applications. Ceramic scaffolds are typically characterized by low elasticity and a hard brittle surface. From a bone perspective, they exhibit excellent biocompatibility to enhance osteoblast differentiation and proliferation. They exhibit excellent biocompatibility due to their chemical and structural similarity to the mineral phase of native bone. Various ceramics have been used in dental and orthopedic surgery to fill bone defects and to coat metallic implant surfaces to improve implant integration with the host bone. While many bone fractures are capable of self-healing, massive bone defects or diseased tissues (i.e., osteoporosis, osteocarcinoma, etc.) often fail to heal properly. Regeneration of damaged or diseased skeletal tissues presents a significant challenge.

Collagen and HA are major constituents of bone. It is appropriate that it be considered as a scaffold material for bone regeneration. Collagen itself has relatively poor mechanical properties. Compressive and tensile mechanical properties of collagen scaffolds can be improved through physical and chemical cross-linking methods. Natural bone is made up of collagen fiber and inorganic HA. The collagen fibrils are coated along their length by HA microcrystals [14].

Research at Tatung University, Taipei, confirmed that the organization of the HA on the collagen has a strong influence on the physical properties and developed a formulation that appears to duplicate this conformation *in vitro* [15]. In their method, type I collagen is dissolved in hydrochloric acid and then precipitated with a calcium hydroxide solution followed by a phosphoric acid. The precipitation was collected as the HA/collagen composite. It was freeze-dried to form a three-dimensional porous scaffold.

In as little as 24 h, crystallinity began to develop in the HA as evidenced by X-ray diffraction (Table 3.6).

In addition to the increased stiffness of the composite, the HA/collagen composite scaffold showed open and interconnected porous structure as observed microscopically. The pore size was in the range of 200–300 μm, which met the requirements to serve as a bone scaffold. The pores appeared to be interconnected and thus would be expected to be able to promote cell growth.

TCP has also been used extensively as a bone substitute due to its similarity to the mineral composition of human bone, excellent biocompatibility, and osteoconductivity [16, 17]. In addition, TCP exhibits a moderate degradation rate to match the rate of osteogenesis [18]. Furthermore, TCP is able to support the attachment, proliferation, and differentiation of various seed cells, including osteoblasts, and adipose- and bone-marrow-derived stem cells [19].

A study at Fudan University, Shanghai, investigated the properties of a construct composed of stem cells attached to a porous TCP scaffold via biotin–avidin bridging [20]. We will discuss this technology again in the next chapter with the focus on immobilization of stem cells. Our discussion here concerns

Table 3.6 The compressive strength of HA/Col composite scaffold.

Sample	Compressive strength (kPa)
Tatung method at 24 h	350
Tatung method at 3 h	123
Simple mixing	36

Source: Rho *et al.* [14]. Reproduced with permission of Elsevier.

the development of the scaffold. Both topics deal with a scaffold for repairing bone defects, and therefore we are concerned with analyzing the porosity, compressive strength, *in vitro* biocompatibility, and *in vivo* bone-forming capacity in a rabbit model.

The porous TCP scaffold was prepared by mixing TCP powder with two particle size grades of ammonium chloride (NH_4Cl). Diameters of 150–200 μm and 60–100 μm were selected as the macroporous and microporous pore-forming agents in NH_4Cl, respectively. The TCP powder and the macroporous pore-forming agent and 0.4 g microporous agent were transferred into a circular mold and subsequently processed by compression. The material was then slowly heated to 400°C and kept there for 1 h. The temperature was then raised to 1100°C. The temperature was kept at 1100°C for 4 h and then decreased to 900°C. The temperature was kept at 900°C for 2 h and then allowed to cool to room temperature (RT) naturally.

The following analyses were performed:

- The porosity of the TCP scaffold was measured using Archimedes' principle.
- The compressive strength of the TCP scaffold was measured.
- The cytotoxicity of β-TCP was also determined.

The scaffold was then prepared for colonization by the stem cells by soaking the sterilized TCP in an avidinylation reagent. A platelet-rich plasma (PRP) was added to the scaffold as well. The stem cells were biotinylated using 3-sulfo-NHS-biotin. This technique is discussed in the chapter on immobilization. Stem cells were seeded onto the TCP/PRP scaffold (with controls of untreated scaffolds and cells) and were incubated at 37°C. Cell counts were taken periodically over a 7-day period. The scaffolds then were washed and prepared for analysis and implantation.

The scaffolds were implanted into bone defects in the mandibles of rabbits. The wounds were monitored by computerized tomography for 4 weeks at which time the animals were sacrificed and the implants examined.

The porous TCP scaffold had a high porosity and suitable pore size, and the utilization of an avidin–biotin binding system increased the adherence of the cells to the scaffold. The addition of PRP increased the rate of new bone formation and improved the quality of bone healing.

Of course, calcium phosphates are not the only ceramic material of clinical interest. The following story is anecdotal but I think significant. As such, it could offer a route to a biodurable scaffold. Soon after we joined the Hypol group W. R. Grace, we found a sample in our desk. We recognized as a fully reticulated ceramic. It was architecturally identical to reticulated PU foam. So much so that we were certain that the foam was made by coating a reticulated foam with a ceramic slurry, dried and fired to burn off the PU and then sintered at high temperature. No one in the organization knew where the sample came from, and so there was no provenance to confirm the method of fabrication. That was considered the likely process, however.

The remarkable property of the material, other than the reticulated architecture, was that when we tried to use a hacksaw to cut the sample, it dulled the saw with no visible effect.

We surmised that the material was to be used in some catalytic process. For instance, ethylene oxide is made using a silver catalyst on ceramic scaffolds in a continuous process. Ethylene oxide is the raw material for producing ethylene glycol and polymers thereof.

We interject it here as an alternative to fabricated scaffolds of biodurable materials, specifically titanium and other metals and composites.

Metals

Scaffolds for the repair of bone defects need to provide sufficient mechanical stability to the surrounding tissue. This is a concern when dealing with open-porous scaffolds or surface coatings. Of course, the bone needs to be able to penetrate into the scaffold.

Metal foams are lightweight and stiff, and therefore they are among the materials of interest for bone repair applications. The scaffold must, of course, have most of the characteristics of the other scaffolds we have discussed. Related directly to bone repair scaffolds, Singh *et al.* [21] listed many of the characteristics to which we are all familiar:

- Biocompatibility
- High porosity
- Interconnected pore structure
- Sufficient space for proliferation of new bone tissue
- Provisions for cell attachment
- Transport of body fluids
- Pore size range, usually 150–500 μm
- Specific to skeletal applications, appropriate mechanical properties for loadbearing

Natural tissue is still considered the best material; however, grafting of autologous material has inherent problems [22]. Therefore, artificial materials are being investigated. Among them are metallic materials and alloys. Titanium in particular has performed well in clinical applications, is commonly available, and can be manufactured in a wide range of forms, including foams.

We will review two projects that use titanium foams produced by two methods.

The first of those is referred to as "space holder." We will find that this technique, by other names, is a common method to develop pore structure in otherwise solid materials. The second method uses sintering techniques with the raw material in a powder form.

Titanium as we said is an accepted material in biomedical application, but when converted to foam, it lacks the compressive strength needed for orthopedic use. To improve the mechanical, chemical, and biological properties, composites using ceramic materials are considered [23]. In this regard, HA would be an obvious choice. The ceramic coatings in general, however, are incompatible with titanium and as a result often flake off. Therefore, the composite materials containing metal and bioceramics must address this problem before clinical use is considered.

In a recent report the authors developed titanium–bioceramic composites with controlled porosity and surface morphology [24]. Corrosion resistance and biocompatibility were also improved. They investigated a composite material described as Ti-Bioglass-Ag.

After milling, the powders were mixed with two particle sizes of table sugar and pressure at 1000 MPa. The sugar was washed out with water. After drying, the composite was fired at 1300°C.

The process yielded a sample with a porosity of 70%. Microscopic analysis showed that the material had a bimodal pore size distribution as a result of the bimodal particle sizes of the sugar. The average pore diameter of about 0.3–1.1 mm with micropores of several microns, and the cells on the Ti-Bioglass-Ag foam were observed to be well dispersed inside the pores.

The compression strength and stiffness for the foam were 1.5 and 34 MPa, respectively. Lastly, the cytotoxicity (against fibroblast) was improved over titanium.

The use of additive manufacturing processes offers the possibility to fabricate open-porous bone scaffolds in a wide range of architectures. The choice of materials dictates the fusion method and the mechanical properties.

In Wieding, scaffolds were made of a titanium alloy by an additive manufacturing process [25]. The scaffolds were fabricated by a selective laser-melting process from titanium powder. Defined pore geometry and porosities of approximately 70% were produced. Three different designs were created of varying structural shapes and strut orientations. The height and diameter of the samples were 14.8 and 4.0 mm, respectively. The first two designs exhibited struts with vertical rectangular cross section. The distance between the two layers was 1.3 mm, and the pore size was 800 × 800 μm. The struts for the third scaffold were diagonal with a circular cross section with a diameter of approximately 300 μm. The distance between the layers was 1.2 mm, and the pore size was approximately 550 × 550 μm. The cell investigations were performed on only one geometric scaffold type. They concluded that it is possible that cell behavior could vary between pores of different geometric shapes.

To determine the viability of human osteoblasts within the scaffold, cells were seeded onto the scaffolds. The cells formed numerous cell connections, which resulted in a densely populated surface on both planes. The osteoblasts were found to be compatible with the porous titanium scaffolds in a static cell culture. The synthesis of pro-collagen type I increased. The scaffold macropores were settled by cells, although the pore size prevented an overgrowing of cells.

The authors concluded that, using selected laser melting in an additive manufacturing process, one could produce appropriate open-porous bone scaffolds. Using titanium in this case, mechanical load on the bone can be controlled and lead to the stimulation for bone regeneration.

Polymer Scaffolds

While ceramics, metals, and composites thereof seem to be the materials of choice for bone repair, by far the most common material for general tissue applications are polymers, thermoplastics, and thermosets. Within this category are both biodegradable and biodurable versions. We have expressed our opinions in this regard, but this review treats them without differentiation.

Polymers intended for tissue engineering applications will have to meet the mechanical properties as we mentioned earlier. In general, there are four types of polymer-based materials under investigation, natural polymers, synthetic polymers, hydrogels, and composites. Of the group, synthetic polymers have the advantage over natural polymers because of their mechanical properties and ability to take on more or less natural architectures (foam, for instance). They are comparatively more predictive and reproducible.

The most extensively used synthetic polymers are poly(glycolic acid) (PGA), poly(lactic acid) (PLA), polycaprolactone (PCL), polyethylene glycol, and various PUs.

Poly(lactic Acid)

PLA is used widely in medical applications and it has been approved by the FDA for implantation in the human body. Based on available data to date, the duration of degradation can be ranged from 12 months to over 2 years. PLA can have a relatively low tensile strength and is primarily used as a nonwoven mesh for tissue engineering applications. Degradation products of these materials reduce local pH, accelerate the polyester degradation rate, and induce inflammatory reaction.

Poly(glycolic Acid)

PGA is an aliphatic polyester. It has a highly crystalline structure with a high melting point and low solubility in organic solvents. The *in vivo* degradation period can be from 4 to 12 months. The degradation products of PGA are resorbed by the body. PGA has a relatively high tensile strength and elasticity. The material is generally formed into a mesh for cartilage tissue engineering purposes. Both PLA and PGA are considered too weak for high porosity scaffolds.

Polycaprolactone

PCL degradation can be as long as 24 months. To increase the rate, it is copolymerized with more fragile polymers. Its biocompatibility, degradation, and mechanical strength characteristics are suitable for orthopedic applications.

These materials are of interest because they are biodegradable. Their mechanical properties are secondary to that. Converting them into three-dimensional scaffolds is problematic. For true scaffold applications, it is the thermoplastics with which we are all familiar are more appropriate. Among those are polyethylene, polypropylene, polyvinylchloride, and, of course, PUs.

It serves the purpose of this section to mention them and defer to the next chapter to illustrate how they are used. It is through the immobilization that converts these otherwise inert thermoplastics into biocompatible scaffolds for cell propagation and attachment of bioactive molecules.

Polyurethanes

PUs play a unique role in the catalog of polymers considered for scaffolding materials. It is the only material that is, or can be, made directly into a three-dimensional architecture, that is, a foam. Furthermore by careful formulation the foam can be open-celled and therefore permeable. With an additional process it can be fully reticulated. It can be either hydrophobic for strength or hydrophilic for biocompatibility. Lastly for this discussion it can be used for immobilization of cell or biomolecules without a separate activation step. This last point is, of course, the subject of the next chapter.

In an earlier chapter we described PUs as a system, not a molecule. While many olefin-type polymers come in several forms, PU is composed of two different molecules, each having a specific effect on the resultant polymer. It is the choice of those components that in large part defines the mechanical properties. In addition, the simple addition of a little water changes the architecture from elastomer to foam.

Focusing on scaffolds, however, we see an excellent library of product designs using the chemistry. The first examples will be cited in several situations. It is in part appropriate to our liver model.

In a recent study, a PU was prepared for culturing chorionic mesenchymal cells [26]. A polyether polyol and methylene diphenyl diisocyanate (MDI) were reacted to make a prepolymer. Water was added, blended, and poured into a mold. The mold was closed and the foam developed at RT. After curing, the PU foam was immersed in absolute ethanol. It was dried in air at RT.

Porosity, average pores size, and pores size distribution were evaluated. The specimens were rotated at 180° around the long axis of the sample. Tridimensional reconstruction of the internal pore morphology was carried out. Scaffold porosity was then calculated from these data. The samples were immersed in Eagle's minimum essential medium.

Table 3.7 Analysis of foam.

Density (g/cm^3)	Average pore diameter (μm)	Porosity/void volume (%)
0.127	268	90

Source: Bertoldi *et al.* [26]. Reproduced with permission of Springer.

Isolated cells from human placenta were suspended in culture medium. A chorionic mesenchymal cell suspension was seeded onto each foam specimen, placed in an incubator, and maintained under static culture conditions for 20 days. As an internal control, cells were cultured on polystyrene wells at the same seeding density.

After 20 days, the PU foam specimens were washed in ethyl alcohol and air-dried. A three-dimensional model generated by micro-CT analysis demonstrated a homogeneous morphology of the foam and regular pore size, shape, and distribution. Micro-CT analyses also allowed the researchers to investigate the average pore size distribution; most of the foam pores have a diameter size in the range between 150 and 400 μm. The pore size, pore volume fraction, and resulting surface area available for cell attachment are well controlled with the proposed scaffold-manufacturing method. The foaming process allows the production of scaffolds with homogeneous structure and morphological properties (Table 3.7).

The authors concluded that the scaffold had good morphological properties. The foam made the material a valid scaffold to support adhesion and differentiation of chorionic mesenchymal cells from human placenta into osteoblasts. Due to the ability to stimulate cell adhesion, scaffold colonization, and osteoblast differentiation, these foamed PU scaffolds appear good candidate as bone graft.

The union of chitosan and PU has an influence in the morphology of the composite. This can be tuned in a way that the scaffold maintains the form desired and the indispensable space for the formation of new tissue via proliferation and differentiation of cells or secretion of its ECM. Chitosan and blends show greater porosity than pure PU; therefore the union of chitosan and PU has an influence in the morphology of the composite. To investigate this [27], PCL diol was dissolved in dioxane and hexane as a cosolvent. Diisocyanatohexane was added with a dibutyltin catalyst. Various amounts of chitosan was added and stirred to obtain different homogeneous mixtures. The composites were then cured at 60°C under reduced pressure for 48 h.

Rat calvaria osteoblast cells were cultured onto the scaffolds. They were seeded at a density of 50 000 cells per sample and then placed in an incubator at 37°C with 5% CO_2 and humidified air. The samples were evaluated for cell surface interactions such as adhesion, proliferation, cell viability, and spreading (Figure 3.3).

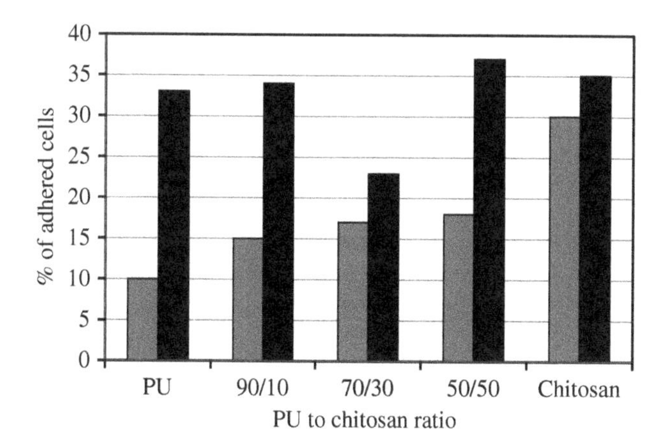

Figure 3.3 Adhesion of osteoblasts to the chitosan/PU scaffolds. *Source:* Imelda Olivas-Armendariz *et al.* [27]. http://www.scirp.org/journal/PaperInformation.aspx?paperID=23807. Licensed under CC BY 4.0.

According to the authors, the PU–chitosan composites were found to have high and interconnected porosity. The temperature and polymer concentration influence the morphological properties of the materials.

In another study, two PU foams were prepared differing in their morphological and mechanical properties [28]. A polyol mixture, water, and Fe-acetylacetonate were added and mixed with a mechanical stirrer. An MDI prepolymer was then added while stirring. The reaction mixture was stirred for a few minutes and then poured into a mold. The mold was closed and the expanding reaction was allowed to take place at RT. One of the foams was made by placing 100 g in a mold, while the other had 90 g. Each foam was extracted from the mold after 72 h, and the skin that forms at the mold surfaces was removed. Not mentioned in the paper was a mold release agent. Its use is typically recommended as the expanding foam penetrates the microstructure of the mold, thereby reducing the useful life by fouling. Many commercial mold release agents are available. In this case the technicians removed the skin.

Density was analyzed according to the European Standard EN ISO 845 by weighing and measuring the specimens after a pretreatment. Porosity, expressed as a percentage of open pores, was evaluated according to EN ISO 4590. The average pore size was analyzed according to ASTM D 3576-94. The morphological investigation was performed by scanning electron microscopy. Compressive mechanical tests were performed under both dry and wet conditions.

Foam EC-1 showed homogeneous pore sizes and distribution, whereas a wide pore size distribution for EC-2 foam was detected with two distinct main ranges of pore dimensions. In contrast, only one class of pore was observed for EC-1. Data of the morphological and physical characterization are presented here (Table 3.8).

Table 3.8 Analysis of the foams.

Sample	Density (g/cm^3)	Void volume (%)	Pore size (μm)
EC-1	0.200	35	691
EC-2	0.200	74	935

Source: Zanetta *et al.* [28]. Reproduced with permission of Elsevier.

According to the authors, both foams showed a similar density. Given that 90 g of EC-2 was poured into mold as opposed to 100 g of EC-1, it is somewhat surprising that the density was the same. However, the larger pore size and void volume is as would be predicted.

Compressive mechanical parameters of the foams were examined. Under dry condition, both PU foams showed similar mechanical properties. This supports the density numbers, but would not be expected based on the pore size.

Under wet condition, a dramatic decrease in the mechanical properties was detected, though a similar behavior was maintained by both foams. This is to be expected. The authors concluded that PUs represent a good choice due to their versatility and the possibility of obtaining a wide range of structures and properties.

In the final example, it is showed how the architecture of PUs makes them a useful scaffold. In this case we examine the research by scientists using PU as a scaffold for cells. While the papers deal with the activity of the composite, structural appropriateness is implied. Whole cells can be immobilized on PU foams by any of several methods. The techniques are described in the next chapter, but for this discussion we wanted to catalog the breadth of research in this area.

With few exceptions the method for immobilization of cells is described in the following example [29]. An *Escherichia coli* strain was grown under aerobic conditions at 37°C. After 24 h the cells were harvested and centrifuged to produce a wet paste. This was mixed with one part of a hydrophilic prepolymer to form an "*E. coli* foam." The foam was cut into 0.5 cm pieces. Though an effective procedure, it is limited from a practical point of view. Firstly this encapsulation technique and therefore activity is diffusion controlled. Secondly the foam produced by this technique is less than ideal with regard to cell architecture. The use of surfactants would have increased mass transport. Even under the best formulation conditions, cutting the foam into pieces was necessary, and even then the metabolism most likely occurred at the surface. Our only point was to illustrate the ease of immobilization.

With that we urge you to review the following catalog of the immobilization of cells on PU chemistry. Many of these projects were selected from a review paper [30] (Table 3.9).

Table 3.9 Cells immobilized on polyurethane.

Organism	Purpose	Reference
Rhizopus arrhizus	Fumaric acid production	M. Petruccioli, E. Angiani, and F. Federici, *Process Biochem.*, **31,** 463 (1996)
Acetobacter aceti bacteria	Vinegar production	I. De Ory, L.E. Romero, and D. Cantero, *Process Biochem.*, **39,** 547 (2004)
Pseudomonas sp. strain	Naphthalene degradation	S. Manohar, C.K. Kim, and T.B. Karegoudar, *Appl. Microbiol. Biotechnol.*, **55,** 311 (2001)
Phanerochaete chrysosporium mycelium	Decolorization of raw sugar	C. Guimarćes, P. Porto, R. Oliveira, and M. Mota, *Process Biochem.*, **40,** 535 (2005)
Phanerochaete chrysosporium mycelia cells	Lignin peroxidase production	S.-S. Shim and K. Kawamoto, *Water Res.*, **36,** 4445 (2002)
Phanerochaete chrysosporium mycelia cells	Manganese peroxidase production in a packed bed	M.T. Moreira, G. Feijoo, C. Palma, and J.M. Lema, *Biotechnol. Bioeng.*, **56,** 130 (1997)
Yarrowia lipolytica yeast cells	Adsorption and degradation of oil	Y.-S. Oh, J. Maeng, and S.-J. Kim, *Appl. Microb. Biotechnol.*, **54,** 418 (2000)
Mycelia of *Aspergillus niger*	Citric acid production in a rotating reactor	W. Jianlong, *Bioresour. Technol.*, **75,** 245 (2000)
Phanerochaete chrysosporium	Degradation of chlorophenols	S.H. Choi, S.H. Moon, and M.B. Gu, *J. Chem. Technol. Biotechnol.*, 77, 999 (2002)
Trametes versicolor	Treatment of kraft paper mill effluent	S. Pallerla and R.P. Chambers, *Tappi J.*, **79,** 156 (1996)
Aspergillus niger cells	Production of gluconic acid	R. Mukhopadhyay, S. Chatterjee, B.P. Chatterjee, P.C. Banerjee, and A.K. Guha, *Int. Dairy J.*, **15,** 299 (2005)
Municipal waste	Degradation of toluene and H_2S	Our research reported here

Source: Adapted from Romaskevicm *et al.* [30].

The "Ideal" Scaffold

With this section we began with what might be considered the ultimate in scaffold design. Recognizing that pore size, surface area, and interconnectedness are among the critical characteristics of a scaffold for clinical remediation, the following might be considered as the ideal in all three. If it were just those factors, it would still be a useful exercise. It is important, however, that we recognize other factors. Among them are tensile, compressive strength, and stiffness as we discussed. The ideal scaffold must be able to adjust to cell-specific requirements. The architecture, however, is the concern of this section.

We will begin the topic of architecture with what we feel represents the ideal structure. We mean this in the sense that it is the highest level possible of void volume, internal surface area, and interconnectedness.

If you will allow us to use a metaphor, we will describe a building with multiple floors. The workspaces are "open concept." That is, they are more like chambers than rooms. We are not concerned about what is in those rooms. They are empty and featureless. Your decision is to evaluate the building to see how it would meet your needs. We will give examples of how the dividers could be decorated, but we will leave that to the next chapter.

We could continue the metaphor, but we think you get the point. What we are describing is a blank within which you will build your device, be it hepatic cells, bacterial, algae, osteoblasts, molds, smooth muscle cells, enzymes, antibodies, or molds. The architecture of the scaffold is the same. Three-dimensionality is a given. If there is any consensus in the field, it is this.

The following list of requirements characterizes an appropriate scaffold, regardless of the application. It is not our intent to simply mention them, but through the discussion we can begin to quantify each of the characteristics. This is an important step for two reasons. Interconnected cells is an example. We all know that it is necessary but the degree to which a structure is interconnected will allow us to compare technologies, and almost as important it would be the first step in building a quality system.

Following the list of characteristics and explanation, we will summarize with a table of the properties that might be considered an ideal scaffold. Within the table are ranges of course, but the intent is to begin the process of determining the most effective scaffold design:

1) The first aspect to be considered is pore size. Many papers, some discussed in the earlier part of this chapter, specified what the ideal pore size range should be.
2) In order to achieve uniform flow through the ideal scaffold, not only does the pore size have to be defined but also the distribution of pore sizes.
3) Void volume has many important aspects. It is the space in a scaffold that can be colonized. In all applications, void volume has a major impact on mass transport through the device.
4) Interconnected cells are required for both spreading of the cell colony and also for efficient perfusion of fluids flowing through it. Related to this is the uniformity of pore size that affects the uniformity of colonization.

5) In as much as cell colonies need a surface on which to grow and spread, high surface area per unit volume would appear to be desirable.
6) The stiffness of the scaffold was discussed earlier. In a general sense, therefore, there must be provisions for adapting the scaffold to specific cell-determined physical requirements.
7) While the surface chemistry is important to cell attachment and survival, again generally, the surface should be neutral.

If we stopped there, we would not be advancing the science. We will use our prerogative and suggest numbers to each of the bullet points. Again this is an ideal scaffold and should be viewed that way. More importantly it suggests that if we are to design scaffolds, they need to be evaluated in the context of those bullet points. To support this, let us present some illustrative values for each of these factors. We begin with pore size.

Pore Size and Distribution

There are conflicting reports on the optimal scaffold mean pore size required for successful tissue engineering. Nevertheless it is a property that must be included, if for no better reason, than the development of a quality system. It has not been conclusively confirmed, although it seems likely that the ideal pore size is cell-type specific. Still further it may be that the ability to spread has more to do with migration or spreading. The effect of mean pore size was studied in a series of collagen scaffolds with pore sizes ranging from 85 to 325 μM [31]. Osteoblast adhesion and proliferation for up to 7 days of post-seeding was determined. The cell count was highest in scaffolds with the largest pore size. However, an early peak in cell number was also seen in scaffolds with a mean pore size of 120 μm.

This suggests that surface area could play a role on initial cell adhesion, but pore size still plays a dominant role. Improved cell migration provided by scaffolds with pores above 300 μm overcome this effect. It was hypothesized that in addition to adhesion and spreading, larger pores lead to a reduction in cell aggregations along the edges of the scaffolds. In the next chapter we will discuss the distribution of bacterial cells in a small-pore hydrophilic PU. It was found that the bacteria congregated at the edges of the foam did not penetrate the foam cubes. This supports their conclusions. Ultimately the paper finds that scaffolds with a mean pore size of 325 μm were optimal for bone tissue engineering.

Rapid prototyping (RP) techniques have been used for tissue engineering applications. The ability to produce predetermined forms and structures including precise pore size with narrow distribution is essential. In addition, the structures can be considered to have 100% interconnected cells. We will discuss this factor shortly. Nevertheless, a study using this construction technique was aimed at producing scaffolds with pore size gradients [32]. The purpose was to determine if enhancing cell seeding efficiency and control could be achieved by this architecture. Scaffolds were based on blends of starch with poly(ε-caprolactone). Scaffolds were made with pore size range 100–750 μm. The mechanical performance of the scaffolds was characterized using dynamic mechanical analysis and conventional compression testing under wet conditions and characterized using scanning electron microscopy. Osteoblasts were seeded onto the scaffolds. Scaffolds with a distribution of pore size were shown to have intermediate mechanical and morphological properties compared with those with uniform pore size. The pore size gradient scaffolds, however, improved seeding efficiency from approximately 35% in homogeneous scaffolds to approximately 70%, but under static culture conditions.

We want to complete this discussion on pore size and then move on to interconnectedness, but first a word about the term pore size. This may be a reflection on the direction of our research but it deserves some attention. Pore size, to us, defines the sizes of the holes between cavities of a foam structure. Images of almost all of the proposed scaffolds in the literature show broad variations in the holes between cells of foam but report pore size as the diameter of the cavities. In the foam business we differentiate among pores and cavities. We have and will continue to use the conventional use of the term. We thought however it might be useful to take a few lines to use our definition and relate it to the diameter of an idealized cavity. The following is an approximation assuming that the foam structure of a free-rise open-cell foam (like a PU) would be if where a perfect dodecahedron with the pore size as the sides of the polygon. Many consider it, rather than a dodecahedron, a 14-sided polygon. For our purposes, however, 12 sides are close enough. In this case the diameter of the cavities is seen in Figure 3.4.

Returning to the distribution of pore sizes, many of the fabrication methods used to create pores seek to control this factor as reported previously. Several attempt to be bimodal as we discussed in the materials section. When scaffolds are made by any of the cited techniques except the rapid prototyping method, despite our best efforts, control is problematic. Remember, when the idea scaffold is developed, it will have to be done in a reproducible manner. Keep in mind that the body makes a physical scaffold (ECM) using the cells themselves as a temporary scaffold. Despite our best efforts, we cannot be that precise. However, we will conclude this chapter with what we feel is a starting point. Let us conclude this section by saying that if cells need a specific pore size range, then pores outside the range will not be used. From a flow point of view, a high percentage of small pores cause the flow to approach a tortuous path. One would want a uniform flow through the structure. This is only possible with a narrow pore distribution.

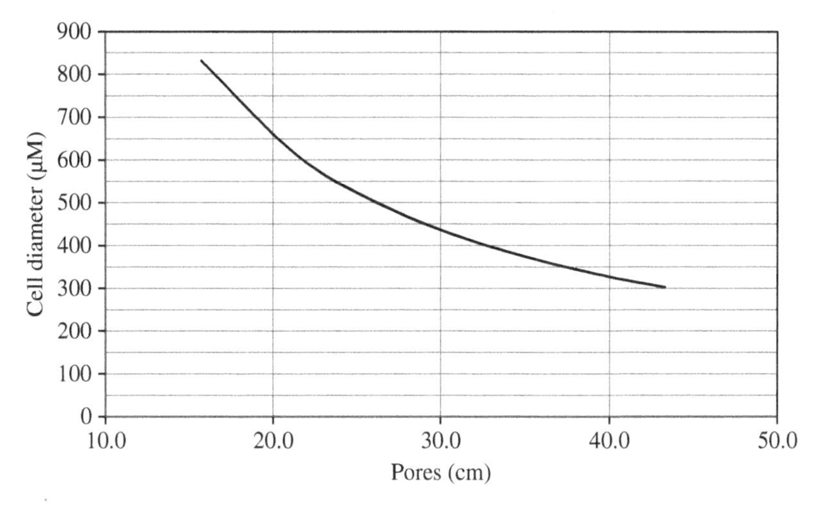

Figure 3.4 Estimate of cell diameter calculated as a dodecahedron.

Void Volume

This is probably the best known and understood of the properties of a scaffold. It is in many ways the most important. It is the reason foam structures are the architecture of most clinically studied materials, which is obvious even though there is no consensus as to what the ideal valve might be. Many of the reverse lost wax procedures yield around 70%.

Karageogiou and Kaplan discussed void volume and pore size as they relate to bone repair [33]. Lower void volume stimulates osteogenesis by suppressing cell proliferation and forcing cell aggregation, while higher void volume and pore size results in greater bone in-growth. However, this results in diminished mechanical properties, thus setting up unofficial limits to pore size and void volume. If we add interconnected cells, we begin to identify an appropriate scaffold.

In water treatment applications, if the bacteria use the pollutants as a carbon source, the buildup of biomass needs the space to grow. High void volume increases the time between cleanouts. One cannot rely exclusively on void volume as the determinant of efficacy. Pore size and void volume in combination are necessary.

An interesting approach is not found in medical or environmental research, but rather in the filtration of molted metal and scaffolds for industrial catalytic processes. Nevertheless it has an interesting potential with regard to void volume of scaffolds. We will cite the work of Louis Wood again when we discuss the immobilization of enzymes. He pioneered many of the applications of hydrophilic PUs as a researcher at the Columbia Research Center of the W. R.

Grace Co. One of those projects is appropriate to this discussion. In a 1974 patent he disclosed a ceramic foam structure prepared by reacting an isocyanate-capped polyoxyethylene polyol reactant with large amounts of water containing a ceramic material [34]. The resultant foams were heated to decompose the carrier foam and sinter the ceramic particles. The result was a rigid ceramic foam structure. The properties of that "ceramic foam structure" are what concerns us here. Other researchers have used this technique and derivatives to produce ceramic foams, and those studies address architecture more completely.

We will begin, however, with the preparation of a ceramic foam as described by Wood. What eventually became Hypol® 2000 was mixed with a slurry of a magnesium iron aluminum silicate ceramic in water. A uniform open-cell structure containing the ceramic resulted. The foam was dried at 130°C. Firing of the dry foamed structure was effected in an air atmosphere by heating at a rate of 150°C/h until a temperature of 1500°C was reached. Firing was continued at 1500°C for 2 h. A sintered ceramic foam resulted, which was free of organic residue and retained the original foamed configuration.

Jumping ahead 20 or so years, we see the work of Minnear [35]. He described a method for forming a porous body of a metal. An aqueous slurry of molybdenum powder containing surfactants was mixed with Hypol 2000. It was allowed to expand freely. The foamed product was dried and then fired at 1400°C in a reducing atmosphere. The resultant foam had a density of 1.33 g/cm^3. The density of molybdenum is 10.22 g/cm^3. Thus the void volume was 87%.

In Peng *et al.* [36], we see improvements in the process, and, more importantly, the architecture is described in detail. While their interest was not in medical and environmental applications, the discussion of the structure of the resultant foam is appropriate. We introduced the difference between the holes that connect the cavities. This aspect is more thoroughly described. The ratio of the window size to the cell size was examined as a useful parameter for characterizing the geometry of the foam and is related to the degree of reticulation.

It is important to describe their focus. Reticulated ceramics are highly porous materials that are used in applications where fluid transport in the microstructure is required. These include molten metal filtration, hot gas filtration, and catalyst supports. These foams are used in aerospace, electronics, and chemical processing.

As we described previously, the typical technique for porous ceramics is based on aqueous media and a polymer binder. The technique is known as "gelcasting."

Their procedure was the same as described by Wood except that they used what is referred to as a two-part PU system, not Hypol. The ceramic powder was predominantly a-alumina and it was used dry (i.e., not as an aqueous slurry).

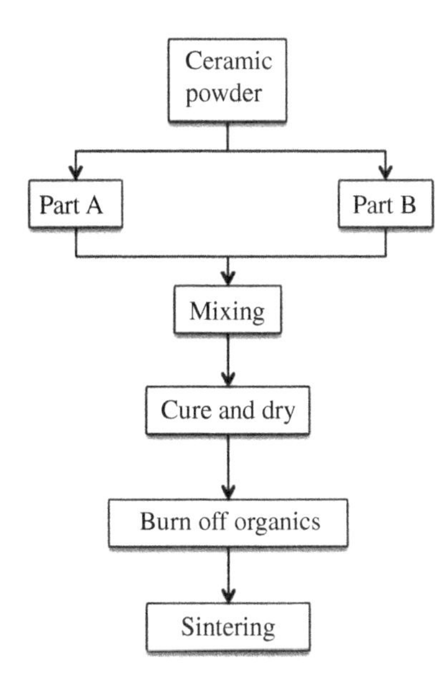

Figure 3.5 Flow sheet of the process to make ceramic foam.

PU is considered a prepolymer, but in a different form. The so-called part A is the isocyanate, and part B is composed of a mixture of polyols, catalysts, cell stabilizers, blowing agents, and other additives. By analogy, Hypol would be part A and water, surfactants, and the ceramic are part B. The chemical nature of the PU was not mentioned, but in as much as powdered ceramic was added directly to part A and part B, we assume it was hydrophobic, as most two-part PUs are. In any case the isocyanate was an MDI. Other than that, the procedures are similar to the Wood patent. The sequence of events is shown in Figure 3.5.

It should be noted that two-part PU systems are commercially available in formulations that produce a broad range of products from foams to elastomers.

For this study, blanks were prepared for comparison at various stages of the process. Both filled and unfilled foams were compared microscopically after curing but before pyrolysis. The micrographs were compared to unfilled foams to show the structure before pyrolysis (Figure 3.6). The main difference is the cell size for the unfilled foam was 590 mm compared with 250 mm for the filled foam. Clearly the mass of the alumina inhibited the free rise of the foam.

Pyrolysis of the foams was done in a temperature-programmable furnace with an airflow of about 4 l/h at a heating rate of 5°C/h to 450°C. Sintering was achieved by heating to 1650°C for 2 h with a ramp of 2°C/min. Problems of agglomerated ceramics were experienced. More significant is a common problem with prepolymers when hand mixing is your only option. The system gels in a few seconds, not giving enough time for thorough mixing. After gelling,

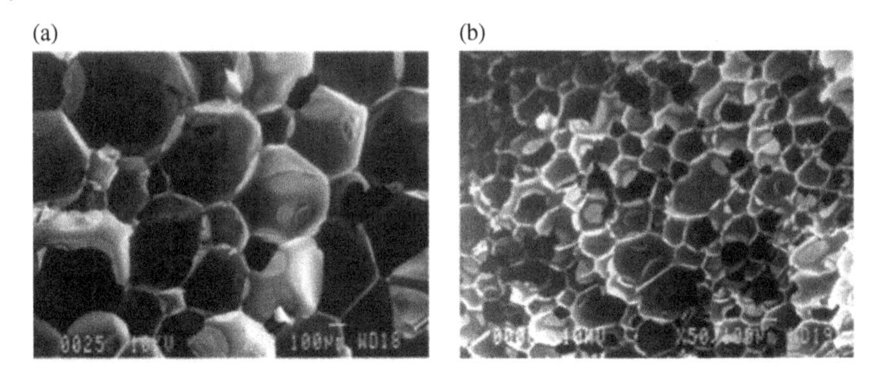

Figure 3.6 Micrographs of unfilled (a) and filled (b) foams before pyrolysis. *Source:* Murphy *et al.* [31]. Reproduced with permission of Taylor & Francis.

more mixing affects the structure of the resultant foam. Nevertheless, the filled PU rose sufficiently to produce foams of 125 kg/m^3, corresponding to a void fraction of 93% QED.

As we have discussed, the foaming process is chaotic in the sense that, once the formulations are mixed, the resultant structure is beyond our control. It is, however, remarkably organized. It is said to be a "natural structure." It has been the subject of mathematical investigation. The current paper continues that. The full details are beyond our purpose but you are encouraged to go to the paper for the discussion. Nevertheless, it is interesting to note that the following mathematical treatment was developed before the invention of PU. The target of the investigation was to determine an ordered structure for bubbles of the same cell size (monodispersed). We will discuss the uniformity of pore sizes in a PU foam, and I think you will be surprised. The most notable investigator in this problem was William Thomson (later to become Lord Kelvin, no relation). He limited his study to systems in which the cell structures were identical. That is to say that the architecture of each cell was the same (as was the size). He examined three specific structures:

- Rhombic dodecahedron composed of four-sided components
- Pentagonal dodecahedron composed of five-sided components
- Orthic tetrakaidecahedron composed of six- and four-sided components

The problem was to develop a structure such that a group of cells fit together without spaces between them.

In 1887, Thomson proposed that the tetrakaidecahedron was the best approximation [37]. It was not perfect, but it was the best fit, that is, the lowest energy. As it happens, constructing a single cell reveals a bowing of some of the components. The structure is built from six squares and eight hexagons and is shown in Figure 3.7.

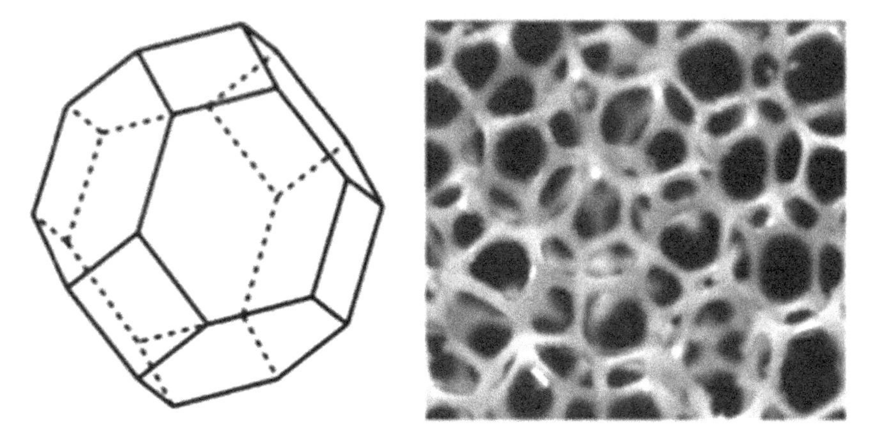

Figure 3.7 A tetrakaidecahedron compared with a reticulated polyurethane.

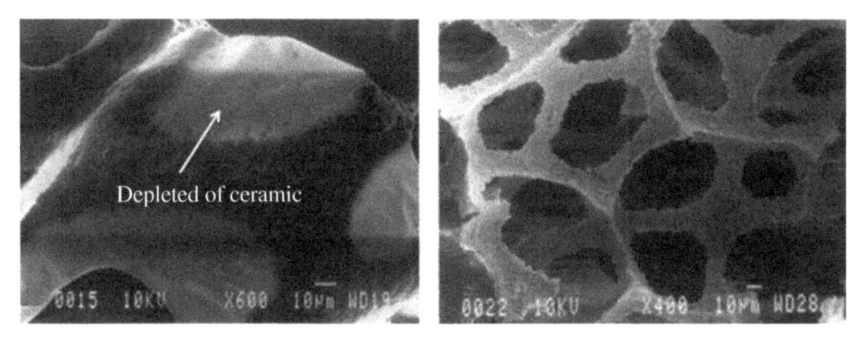

Figure 3.8 Effect of pyrolysis on ceramic-filled foam. Arrow indicates window in the unfired foam. *Source:* Peng *et al.* [36]. Reproduced with permission of Elsevier.

Returning to Peng *et al.*, the microstructure of the polymer–ceramic foam shows that the ceramic powder segregates to the struts. This may be associated with the elongational flow of a thin film of similar thickness to the particle. It is this effect that leads to the reticulation. Because the "windows" have little or no ceramic, the windows are lost upon pyrolysis (Figure 3.8).

While the conversation for our purposes could stop here, the discussion went into great detail to describe the architecture. All we need to know however is described previously. If your research is to produce a stiff high surface area and high void volume foam scaffold, this should certainly be considered an option. Again this is the ultimate in stiffness. We had a sample produced by, we assumed, Wood using a boron nitride ceramic. It could not be cut with anything short of a diamond-tipped saw.

The cell shape is polyhedral and the average cell diameter is 150 mm with a narrow size distribution. This is due to the nature of the foaming process and will be discussed in the succeeding text. All the cells have open windows and the average diameter of the windows was 70 mm.

Minnear [35] characterized the reticulation using a slightly subjective scale related to cell wall thickness and hence void fraction. The cellular structures in the present work suggest that a coordination number of 12 is appropriate. That is, each cell is connected to 12 other cells. Lord Kelvin would have argued 14.

The discussion goes on to describe commercial alumina foams, not reticulated, that are spherical. In terms of measurement, there are numerous techniques, all roughly equivalent. For the most part, comparing densities is sufficient as long as you know the density of the solid material. Various methods based on water displacement are used. Various pycnometer methods have been described. In our work we needed a rapid method to screen candidates. While not appropriate for precise research, it is quick and easy. In our work, we measure the mass rate of water draining from a scaffold. However, this requires relatively large sample sizes.

Interconnectedness

In our review of the literature, interconnectedness appears to be a qualitative measure. As we move forward in scaffold design, it will become important to quantify this important aspect. Proliferation requires spreading and a certain level of openness. That openness is what we need to define.

A study was designed to determine the effect of interconnective pore size on chondrocyte proliferation in a chitosan and PGA scaffold for chondrogenesis [38]. Chondrocytes were seeded onto pre-wetted scaffolds having the following sizes:

1) Pores ≤10 μm in diameter
2) Pores measuring 10–50 μm in diameter
3) Pores measuring 70–120 μm in diameter

The samples were cultured in a rotating bioreactor. Interconnectedness was confirmed microscopically. Chondrocyte proliferation and metabolic activity improved with increasing interconnected pore size.

In our discussion of pore size, we reported on rapid prototyping technique. The study addressed pore size, but it is the nature of the technique to develop 100% interconnected cells. This shifts the discussion. In all other techniques however, pore size and interconnectedness are inseparable to the goal.

While in most cases the property still remains somewhat qualitative. It is clear that interconnectedness must move from qualitative to a measurable property. This was recognized in a paper developed at the Biomechanics Laboratory at the University of Toulouse, Toulouse, France [39]. They developed an approach to improve the measure of the fluid flow in biomaterials with interconnected porosity. An HA having a well-defined porosity was used as a surrogate for typical bone architecture. Fluid velocities within the HA samples were characterized using high-resolution MRI in conjunction with the measurement of global flow and associated permeability. Image analysis permitted computation of local porosity, intra-pore fluid shear, and visualization of flow heterogeneity within the sample. These results may benefit applications in biomaterials for the evaluation of factors influencing bony incorporation in porous scaffolds and on porous implant and bone surfaces.

In what may seem like being offtrack, we think it is nonetheless instructive. It also approaches interconnectedness as a measurable characteristic. Acoustic absorbing foam materials were produced from PU [40]. The pore sizes of the foams were varied from 0.35 to 1.05 mm in diameter. The pore cells were interconnected with open porosity in the range from 16.0 to 88.6%. In other words they were open-cell foams similar to what would be used as a scaffold for colonization. The compressive strength of the foams decreased with increase of interconnected cell ratios. There was a decrease of the strength lower than 92.6% compared with the closed cell sample when interconnected cell ratio was 88.6%. The acoustical efficiencies of PU foams were studied, and the results showed that the porous cell size and interconnected porosity of PU foams had significant influence on the acoustic efficiencies.

The cited reference makes the point that interconnectedness is important and there are ways to measure it. While at this stage efficacy is the driving forces as the technology develops, quality control will require some level of uniformity batch to batch. Thus a quantitative measurement is critical. Not mentioned in the aforementioned discussion, but equally important, is that fluids flowing through a scaffold must have full access to the surfaces. Nevertheless, as we said, it is common to comment on interconnected cells as a feature, but without the degree to which the scaffold meets a minimum requirement.

While the techniques described previously are interesting and perhaps useful, a simpler method is appropriate. Fortunately, we can turn to engineering science for an indirect measure for interconnectedness. Simply put the higher the degree of interconnectedness, the higher the flow rate. Conversely, the pressure it takes to pass a fluid through a vessel at a given flow rate is a measure of interconnectedness. In the industry it is defined as pressure drop. We discussed this in the laboratory practices chapter. A physiological example is the pressure drop across the liver. Portal hypertension is the elevation of hepatic venous pressure drop above 6.7 kPa. It is caused, in part, by increased resistance to passage of blood flow through the liver [41]. High blood pressure, described this way, is a symptom of obstructed flow and it puts a burden on the heart.

In the next chapter, we will describe a scaffold colonized with bacteria for the treatment of contaminated air or water (biotrickling filter). We will explain

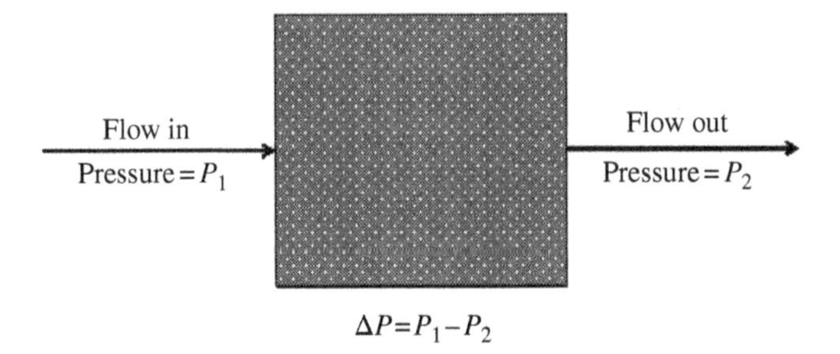

Figure 3.9 Pressure drop across a scaffold.

that in as much as reactions are not instantaneous, the fluid must have a finite time in the filter. That time is in part controlled by the flow rate. The flow rate is, of course, generated by a pump, but depending on certain physical factors, pressure develops at the entrance to the vessel. The required flow rate determines that pressure. Remember, the scaffold is a constant, but the flow rate and therefore the pressure are determined by the required residence time in the filter. A higher degree of interconnectedness requires less pressure.

In engineering practice, the pressure drop is part of the design process (size and power of the pumps) (Figure 3.9). In-use flow rates can be adjusted by pump speed. *In vivo*, however, we don't have that option.

With that as the justification, we propose that medical scaffold developers begin to consider interconnectedness as a variable and begin to quantify it. While liquids should and will eventually be used, it is convenient, and the essential structural aspects are served by using air as the fluid. Using liquids introduces viscosity as a variable and is unnecessary to the goal of identifying structural factors. In Figure 3.10, the pressure drop across a foam serves as an example of how a data set might look and is represented as a function of pore size (not interconnectedness, however). It has nothing to do with the values that a physiologically appropriate scaffold would have. However, it is how an air filter might be evaluated. Note the pressure drop relative to what would be required for blood flow.

Thus far we have discussed pore size, void volume, and interconnectedness. We feel these are the most important factors, but there are other aspects that need to be considered. In many cases, the first three factors cannot be achieved without consideration of the following.

Surface Area

High on the list of desired properties of an appropriate scaffold is surface area. Cells typically need an attachment site to proliferate. Spreading and even cell-to-cell attachment is in part affected by the surface. For most applications,

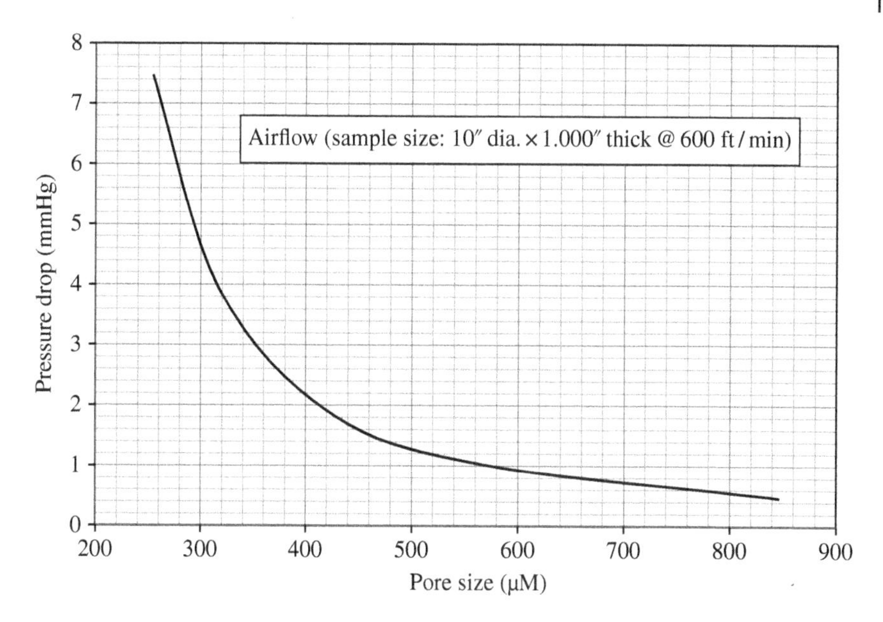

Figure 3.10 Pressure drop across a scaffold at 200 fpm. *Source:* Restated from FXI Corp. Product Literature.

including our liver model, sufficiently high surfaces are clinically effective or efficient for the treatment of contaminated fluids. Surface area of a scaffold and pore size is connected but the relationship is complex. For open-cell materials, partial membranes between cavities are surfaces for colonization. The state of the art does not have an upper limit to surface area as long as the other important factors of pore size and interconnectedness are appropriate.

Having said that, there are methods to measure surface area. Methods such as the BET method, low adsorption methods, and flow methods are used [42]. The methods are appropriate for various types of surfaces, including heterogeneous surfaces, porous solids, clays, and small area surfaces. In practice, like interconnectedness, discussion of this important aspect is assumed to be obvious from porosity and pore size measurements. High internal surface area-to-volume ratios are essential in order to accommodate the large numbers of cells required to replace or restore tissue or organ function(s). For a given material and process, pore size and internal surface area are related. Adjustment of that relationship between both properties has to be determined, all dependent on the application. Besides maximizing the surface area, the morphology and chemistry of the scaffold's surface are important factors that influence cell attachment, migration, and intracellular signaling *in vitro* and cell recruitment and healing at the tissue-scaffold interface *in vivo*.

Mechanical Properties

For the most part, the scaffold will not be physically stressed by the environment. A possible exception is bone and cartilage that appear to be influenced by the elasticity of the scaffold—in this respect tensile and compressive strength and a related property, stiffness. The measurement of these properties is covered in standard ASTM D3574 and discussed in the chapter on analytical procedures.

From an engineering perspective, strength is critically important. When the scaffold is not sufficiently strong, external structures (like a metal frame) are built to support it. As such engineers understand the practical implications of the strengths of materials. We know however that the mechanical properties of a scaffold intended for mammalian cell propagation, to some degree, influence cell differentiation [25]. This brings mechanical properties into our list of factors to be considered. In the extreme, bone replacement technologies show not only the importance but also the difficulty in defining a target. Animal skeletal bones have tensile strengths between 40 and 200 MPa [43]. Nevertheless, measurements of this and other mechanical factors are necessary to advance scaffold science. In this regard, values for both tensile and compressive strengths need to be developed to define a tissue-mechanical relationship.

The following are stylized curves that might be developed on testing equipment used to develop strength measurements according to the aforementioned ASTM standard (Figure 3.11). They measure the strength required to stretch a sample and then compress it.

Surface Chemistry

Surface chemistry is best discussed in the next chapter. While the construction of a scaffold is important, it is insufficient to the task. Any number of materials can be used as long as the requirements discussed previously are used as a goal. It is the surface chemistry that is ignored in that discussion. We know that it is the association of the cells to that surface that ultimately defines efficacy. It is not sufficient to create a neutral surface if one is to make a device appropriate for clinical applications. This is and should be the goal for environmental applications where at least bacteria have evolved to their own method of attachment. For mammalian cells however the surface chemistry must be appropriate. In addition, if this is to be a blood contact device, surface chemistries must be considered. We will discuss this in the next chapter, but for now much of the research around the world has focused on this aspect. It is our goal to combine architecture and surface chemistry. This chapter discusses structure and we will leave chemistry to the next.

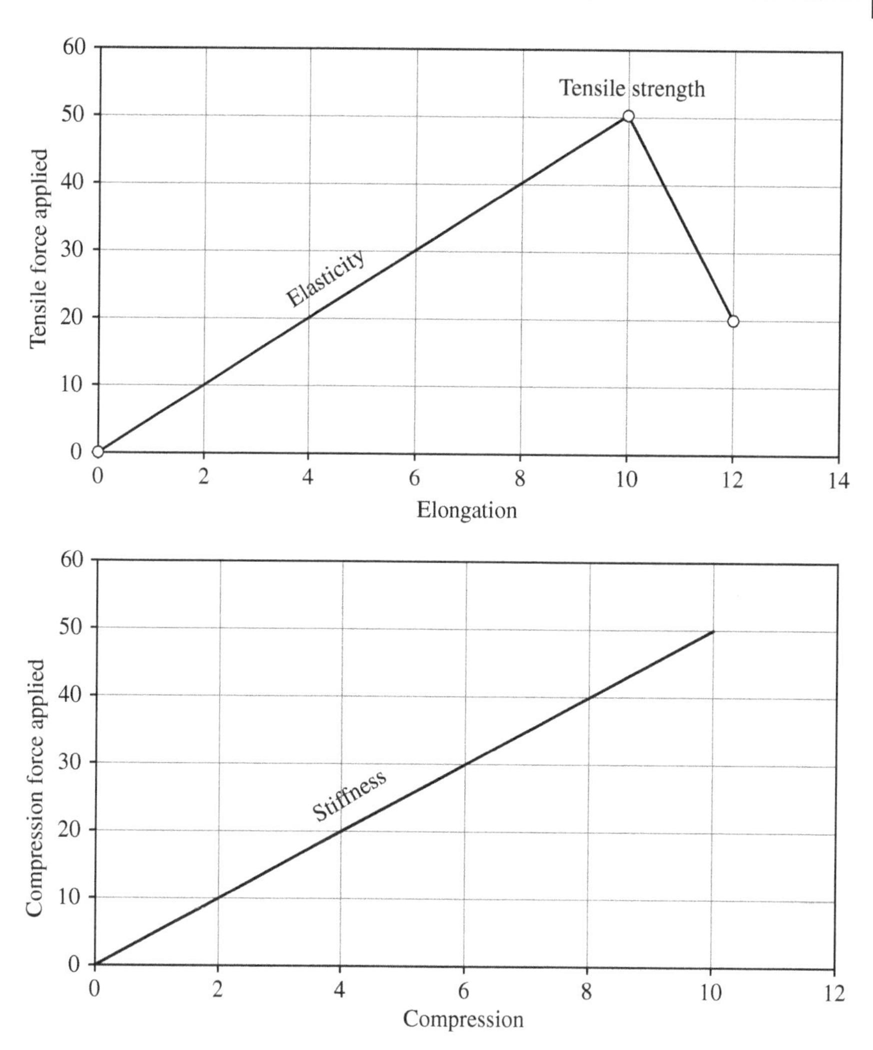

Figure 3.11 Idealized stress–strain curves (tensile on the left, compression on the right).

Specifications of the Ideal Scaffold

Others have assigned certain characteristics to the ideal scaffold. In fairness they established the goal of tissue engineering as an implantable device. In retrospect, however, we should have set functional milestones. Those checkpoints should have had clinical significance to be sure. To not have established a route to the ultimate destination, we have gotten lost in our opinions. Without knowing what a scaffold should look like has resulted in studies around the world with virtually the same objectives. Ranges of values are reported to be sure, but we would be hard pressed to define the best pore size and interconnectedness for our liver model. Factors like interconnectedness are discussed, but not in the context of flow, mass transport, or residence time. Properties like biodegradability have overly influenced the route to a liver support device. While we also consider this the ultimate goal, we have not answered the comparatively simple quests of mass transport. Many researchers have recognized the need for tackling the engineering before the tissue part is developed. We are thinking specifically of the work in Kyushu University. We and others around the world have felt that jumping into an implantable device that would magically disappear when its purpose was fulfilled was possible, but defining the architecture of the scaffold was the immediate task. We went to the moon first and then, we expected to use a moon base to launch the Mars mission.

In the first part of this section, we discussed the important aspects of tissue engineering scaffolds as defined by a broad spectrum of the research community. We will summarize these data to describe the ideal platform. We recognize that this is highly presumptuous, but we do not apologize for what we are about to recommend. It is a view of tissue engineering from a distance. This allows us to see trends and accomplishments that are outside the focus required by day-to-day practice. The complexity of work in any *in vivo* process is consuming. We have had the opportunity to work on engineering projects in which colonization was not only natural but also unavoidable. Accordingly, innovation focused on scaffold.

What we are about to suggest can be taken in two ways. It can be taken as a template on which a scaffold could be developed (which is what we prefer) or as the ultimate of what a scaffold could be. If high surface area, interconnected cell, and so on are the goal, this is what that scaffold would look like. As a template it can serve as a structure upon which a tissue can be built. We have often referred to this as what God used to develop us. That is, hydrogel system built on a skeletal structure.

In the other sense, it can at least be a target to direct scaffold research. While keeping this aspect in mind, we will discuss this ultimate scaffold as a template.

We begin with a micrograph of how the scaffold might look (Figure 3.12).

Many of you will recognize this for what it is, a reticulated hydrophobic PU foam originally developed as an air filter. It is used in many ways including in the fuel tanks of fighter jets and race cars.

Tracing their origin back to air filters, reticulated foams are differentiated by pore size from 1 cm to 250 μm. They are made commercially using a hydrophobic polyol and toluene diisocyanate using the "one-step" process described in the chemistry chapter. Pore size is very carefully controlled. This is followed by the reticulation step described earlier.

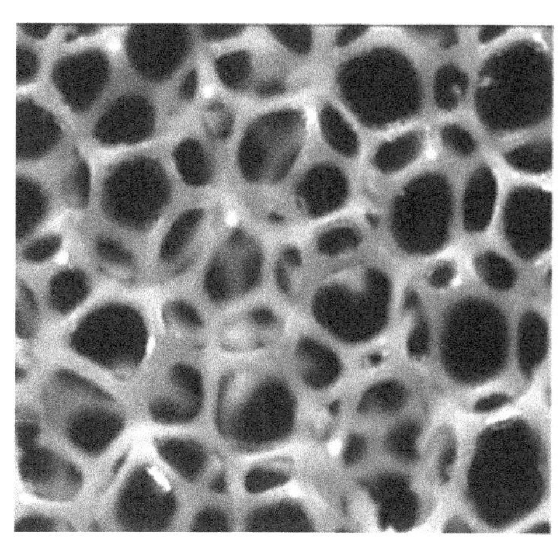

Figure 3.12 An idealized scaffold.

This is not a comprehensive list of manufacturers, but reticulated foams are produced in the United States, by Crest Foam (Moonachie, NJ) and FXI (Media, PA). Recticel in Belgium produces these foams as does Inoac in Japan.

For our purposes it is the basis for the development of a set of specifications. We have compiled them in Table 3.10.

Again whether you consider this a template on which one could build a tissue engineering scaffold or simply as an example of what an ideal scaffold might look like, one has to agree that it is a remarkable material.

It is paradoxical that despite the breadth of the pore sizes, the density and therefore the void volume are unchanged. The pore size distribution around the nominal size approaches monodispersed. As we mentioned in the text, this is an important and yet unexplored characteristic. It logically affects the uniform spreading of the cells and certainly the perfusion of liquids through it. Most important for tissue engineering is the range of pore sizes and the surface area. The smallest pore sizes are within the ranges mentioned earlier (256, 322, and 416 μm, respectively).

An unappreciated factor but nevertheless important is that these foams are available commercially to everyone, with certificates of compliance.

Having said all that, for these foams, which would appear to be an excellent candidate for consideration, there are factors that need to be considered. First for what appears to be the largest community of researchers, they are not biocompatible. Again for our work, this is not a problem. In addition it is the nature of the business that commercial innovation is by an unmet need. In this case there is no current need for a biodegradable reticulated foam. The chemistry of PU includes that potential as we described in the chapter on chemistry. When or if a "market" develops, innovation will begin.

Tables 3.10 The physical of our ideal scaffold.

Nominal pore size (pores/cm)	Min.	Max.	Density (g/cm^3)	Void volume (%)	Surface area (m^2/m^3)
39	31	43	1.2	98	7382
31	28	35	1.9	97	5578
24	22	26	1.9	97	3937
18	16	20	1.9	97	2625
12	10	14	1.9	97	1641
10	8	12	1.9	97	1181
8	6	10	1.9	97	984
4	3	6	1.9	97	492

Nominal pore size (pores/cm)	Surface area (m^2/m^3)	Pressure drop (mmHg) sample size: 10″ dia. × 1.000″ thick at 600 ft/min	Density (g/cm^3)	Void volume (%)
39	7400	7.47	0.02	97
31	5700	3.92	0.03	97
24	3900	1.87	0.03	97
18	2600	1.03	0.03	97
12	1600	0.47	0.03	97
10	1200	0.35	0.03	97
8	1000	0.28	0.03	97
4	490	0.19	0.03	97

Source: Adapted from Kelvin [37].

The surface while arguably biocompatible is hydrophobic and more importantly hemolytic. Work in Japan on a liver support device using a reticulated foam required that only blood plasma contacted the scaffold [44].

Also, as part of the limitations of the material as described thus far, it is accepted that if your product is to repair bone defects, a stiffened perhaps rigid scaffold is required. This is clearly not a property of this foam.

These and other limitations however serve as an introduction to the next chapter. We have described this material as a template. In this respect we can use the very high void volume to develop tissue-specific surfaces. Using a technique known in the ceramic industry as slip casting, we can build a ceramic scaffold with nearly the same architectural characteristics. We can graft other polymer onto the surface to take advantage of specific bioactive chemistries. In summary, it is our opinion that this could be a platform on which we develop a surface that meets the need of a biomedical application. This includes the immobilization of enzymes, antibodies, antigens, and other bioactive molecules.

In conclusion, we have separated our discussion into two parts, architecture and chemistry. It is our opinion that both are necessary for an effective bioscaffold, but neither is sufficient. Engineers know this but have the advantage that the surface attachment of bacteria is an evolutionary property of the organism itself. Bacteria are designed to stick to surfaces. Therefore their research focus has been on the architecture. We can and have learned from their experience. It is now the task to direct our attention to chemistry, the subject of the next chapter. This does not, however, leave the environmental engineers off the hook. Who is to say that the propagation of animal cells cannot be used to address environmental concerns? Certainly, yeast and mold cells are currently used. Algae is of interest in capturing heavy metals including uranium. Thus we continue our theme of creating a multidiscipline forum.

References

1 Recirculated water systems in aquaculture, Muir, J.F., *Recent Advances in Aquaculture*, Ed. Muir, J.F. and Roberts, R.J., Westview Press, London, 1982.

2 Structure Property Relations, Szycher, M., Presented at the *12th Annual Seminar on Advances in Medical-grade Polyurethanes*, Technomics Publishing, Lynnfield, MA, 1998.

3 FXI Corp. Technical Product Function Sheet, FS-998-F-5M. FXI (Foamex Corp), Media, PA.

4 *Biofiltration for Air Pollution Control*, Devinny, J.S., Deshusses, M.A., and Webster, T.S., CRS Press, Boca Raton, 1999.

5 *The ICI Polyurethane Book*, Wood, G. John Wiley & Sons, Ltd, Chichester, 1987.

6 The measurement of liver circulation by means of colloidal disappearance rate, Dobson, E.D., Warner, G.F., Finney, C.B., and Johnson, M.A., *Circulation* VII, 690–695, May 1953.

7 High metabolic function of primary human and porcine hepatocytes in a polyurethane foam/spheroid culture system in plasma from patients with fulminant hepatic failure, Yamashita, Y.-i., Shimada, M., Tsujita, E., Shirabe, K., Ijima, H., Nakazawa, K., *et al.*, *Cell Transplantation* 11, 379–384, 2002.

8 The basement membrane component of biologic scaffolds derived from extracellular matrix, Brown, B., Lindberg, C., Reing, J., Stolz, D.B., and Badylak, S., *Tissue Engineering* 12 (3), 519–526, 2006.

9 Matrix elasticity directs stem cell lineage specification, Engler, A.J., Sen, S., Sweeney, H.L., and Discher, D.E., *Cell* 126, 677–689, August 25, 2006.

10 Cutting the cost of drug development?, Rawlins, M.D., *Nature Reviews Drug Discovery* 3 (4), 306–4, April 2004.
11 Drug induced hepatotoxicity: a comprehensive review, Kshirsagar, A., Vetal, Y., Ashok, P., Bhosle, P., and Ingawale, D., *The Internet Journal of Pharmacology* 7 (1), 2008.
12 Functional differentiation and alveolar morphogenesis of primary mammary cultures on reconstituted basement membrane, Barcellos Hoff, M.H., Aggeler, J., Ram, T.G., and Bissell, M.J., *Development* 105, 223–235, 1989.
13 Modeling tissue-specific signaling and organ function in three dimensions, Schmeichel, K.L. and Bissell, M.J., *Journal of Cell Science* 116, 2377–2388, 2003.
14 Mechanical properties and the hierarchical structure of bone, Rho, J.Y., Kuhn-Spearing, L., and Zioupos, P. *Medical Engineering & Physics* 20, 92–102, 1998.
15 Development and characterization of a bioinspired bone matrix with aligned nanocrystalline hydroxyapatite on collagen nanofibers, Wu, H.-C., Wang, T.-W., Sun, J.-S., Lee, Y.-H., Shen, M.-H., Tsai, Z.-R., *et al.*, *Materials* 9, 198, 2016.
16 Bone augmentation in rat by highly porous β-TCP scaffolds with different open-cell sizes in combination with broblast growth factor, Miyaji, H., Yokoyama, H., Kosen, Y., Nishimura, H., Nakane, K, Tanaka, S, *et al. Journal of Oral Tissue Engineering* 10, 172–181, 2013.
17 Effect of a β-TCP collagen composite bone substitute on healing of drilled bone voids in the distal femoral condyle of rabbits, Zheng, H., Bai, Y., Shih, M.S., Hoffmann, C., Peters, F., Waldner, C., and Hübner, W.D., *Journal of Biomedical Materials Research Part B: Applied Biomaterials* 102, 376–383, 2014.
18 Mechano-chemical synthesis and characterization of nanostructured β-TCP powder, Choi, D. and Kumta, P.N., *Materials Science and Engineering: C* 27, 377–381, 2007.
19 Exploring the application of mesenchymal stem cells in bone repair and regeneration, Griffin, M., Iqbal, S., and Bayat, A., *Journal of Bone and Joint Surgery British* 93, 427–434, 2011.
20 Biotin-avidin mediates the binding of adipose-derived stem cells to a porous β-tricalcium phosphate scaffold: Mandibular regeneration, Feng, Z., Liu, J., Shen, C., Lu, N., Zhang, Y., Yang, Y., and Qi, F., *Experimental and Therapeutic Medicine* 11, 737–746, 2016.
21 Titanium foams for biomedical applications: a review, Singh, R., Lee, P., Dashwood, R., and Lindley, T., *Materials and Technology* 25, 127–136, 2010.
22 Complications after harvesting of autologous bone from the ventral and dorsal iliac crest—a prospective, controlled study, Niedhart, C., Pingsmann, A., Jurgens, C., Marr, A., Blatt, R., and Niethard, F.U., *Zeitschrift für Orthopädie und ihre Grenzgebiete* 141, 481–486, 2003.

23 Fabrication and properties of titanium-hydroxyapatite nanocomposites, Niespodziana, K., Jurczyk, K., Jakubowicz, J., and Jurczyk, M., *Materials Chemistry and Physics* 123, 160–165, 2010.

24 Nanostructured titanium-10 wt% 45S5 bioglass-Ag composite foams for medical applications, Jurczyk, K., Adamek, G., Kubicka, M.M., Jakubowicz, J., and Jurczyk, M., *Materials* 8, 1398–1412, 2015.

25 The effect of structural design on mechanical properties and cellular response of additive manufactured titanium scaffolds, Wieding, J., Jonitz, A., and Bader, R., *Materials* 5, 1336–1347, 2012.

26 Ability of polyurethane foams to support placenta-derived cell adhesion and osteogenic differentiation: preliminary results, Bertoldi, S., Fare, S., Denegri, M., Rossi, D., Haugen, H.D., Parolini, O., and Tanzi, M.C., *Journal of Materials Science: Materials in Medicine* 21, 1005–1011, 2010.

27 In vitro evaluation of polyurethane-chitosan scaffolds for tissue engineering, Imelda Olivas-Armendariz, I., Perla García-Casillas, P., Estradal, A., Martínez-Villafañe, A., De la Rosa, A., and Martínez-Pérez, C.A., *Journal of Biomaterials and Nanobiotechnology* 3, 440–445, 2012.

28 Ability of polyurethane foams to support cell proliferation and the differentiation of MSCs into osteoblasts, Zanetta, M., Quirici, N., Demarosi, F., Tanzi, M.C., Rimondini, L., and Fare, S., *Acta Biomaterialia* 5, 1126–1136, 2009.

29 Immobilization of *Escherichia coli* cells containing aspartase activity with polyurethane and its application for L-aspartic acid production, Fusee, M.C., Swann, W.S., and Calton, C. J., *Applied and Environmental Microbiology* 42 (4), 672–676, October 1981.

30 Application of polyurethane-based materials for immobilization of enzymes and cells: a review, Romaskevicm, T., Budriene, S., Pielichowski, K., and Jan Pielichowski, J., *Chemija* 17 (4), 74–89, 2006.

31 The effect of mean pore size on cell attachment, proliferation and migration in collagen–glycosaminoglycan scaffolds for bone tissue engineering, Murphy, C.M., Haugh, M.G., and O'Brien, F.J., *Biomaterials* 31 (3), 461–466, January 2010.

32 Three-dimensional plotted scaffolds with controlled pore size gradients: Effect of scaffold geometry on mechanical performance and cell seeding efficiency, Sobral, J.M., Caridade, S.G., Sousa, R.A., Mano, J.F., and Reis, R.L., *Acta Biomaterialia* 7 (3), 1009–1018, March 2011.

33 Porosity of 3D biomaterial scaffolds and osteogenesis, Karageogiou, V. and Kaplan, D., *Biomaterials* 26 (27), 5474–5491, September 2005.

34 Method of preparing porous ceramic structures by *ring a polyurethane foam that is impregnated with inorganic materials, Wood, L.L., Messina, P., and Frisch, K.C., US Patent No. 3833386, September 1974.

35 Processing of foamed ceramics, Minnear, W.P., In *Ceramic Transactions 26: Foaming Science and Technology for Ceramics*, Ed. Cima, M.J., pp. 149–156, American Ceramic Society, Westerville, 1992.

36 Microstructure of ceramic foams, Peng, H.X., Fan, Z., Evans, J.R.G., and Busfield, J.J.C., *Journal of the European Ceramic Society* 20 (7), 807–813, 2000.
37 On the division of space with minimum partitional area, Kelvin, L. (Sir William Thomson), *Philosophical Magazine* 24 (151), 503, 1887.
38 Chitosan scaffolds: interconnective pore size and cartilage engineering, Griffon, D.J., Sedighi, M.R., Schaeffer, D.V., Eurell, J.A., and Johnson, A.L., *Acta Biomaterialia* 2 (3), 313–320, May 2006.
39 Use of high-resolution MRI for investigation of fluid flow and global permeability in a material with interconnected porosity, Swider, P., Conroy, M., Pedrono, A., Martell, S., Ambard, D., Soballe, K., and Bechtold, L.E., *Journal of Biomechanics* 40 (9), 2112–2118, 2007.
40 Correlation between the acoustic and porous cell morphology of polyurethane foam: effect of interconnected porosity, Zhang, C., Li, J., Hu, Z., Zhu, F., and Huang, Y., *Materials & Design* 41, 319–325, October 2012.
41 Clinical manifestations of portal hypertension, Al-Busafi, S.A., McNabb-Baltar, J., Farag, A., and Hilzenrat, N., *International Journal of Hepatology* 2012, 203794, 2012.
42 *Surface Area and Porosity Determinations by Physisorption*, Condon, J., Elsevier Science, Amsterdam, 2006.
43 Collagen-hydroyappitite composites for hard tissue repair, Wahl, D.A. and Czernuszka, J.T., *European Cells and Materials* 11, 43–56, 2006.
44 Evaluation of a hybrid artificial liver using a polyurethane foam packed-bed culture system in dogs, Gion, T., Shimada, M., Shirabe, K., Nakazawa, K., Ijima, H., Matsushita, T., *et al.*, *Journal of Surgical Research* 82, 132–136, 1999.

4

Immobilization

Introduction

Before we focus on immobilization, it is important to note that its utility is through scaffolds of high surface area, high void volume, and so on. We are about to describe how we and others have taken full advantage of those attributes to develop biological or chemical devices. We have a prejudice toward flow-through devices, but the application of scaffold technology is much broader. Nevertheless, the primary purpose of this text is to explore flow-through applications. In environmental uses, we are concerned with the flow of water or air. In medical apps, the fluid might be blood. In either case the flow characteristics are described by the qualities of scaffolds defined in the last chapter. We summarized them as architectural. In this chapter, however, we are more concerned with chemistry.

If you will permit a digression, however, the concept of scaffold extends beyond flow-through. The repair of diseased or damaged bone is an important topic for discussion. There are other concepts that may be served by a more or less solid scaffold and therefore do not require the flow of fluids.

By way of example, if you are a baker, the action of yeast on sugar causes bread to rise, creating an edible scaffold. Portland cement concrete and asphaltic concrete use a binder supported by a scaffold of aggregate. The fact that we fill much of the void volume does not change the definition. There is a grade of asphalt concrete called open-graded friction course in which the sizes of the aggregate are mixed and the amount of asphalt is carefully controlled to form a permeable road surface that drains water—clearly a scaffold application. The repair of bone, as we mentioned, is a rigid scaffold fully colonized with osteoblasts.

From another point of view, scaffolds are a subset of the now ubiquitous composite technology. A scaffold material (carbon fiber, wood chips, wood

Polyurethane Immobilization of Cells and Biomolecules: Medical and Environmental Applications, First Edition. T. Thomson.

itself, etc.) is bound together with a binder. Wood is a composite of linear cellulose bound by lignin. In this sense it is not different than a carbon fiber composite.

We have drifted away from the narrow, yet important, discussion of scaffolds in the last chapter. Again, our focus has been on flow-through devices and the examples in the above paragraphs have deviated from the topic. There was a reason for these examples and it had to do with the surface of the scaffold. While the last chapter describes architecture, in this chapter we will focus on the chemistry. Using one of the examples earlier, chemicals are used to modify the surface of the stone aggregate to improve the bonding with the cement.

In this chapter we will focus in a general way on modifying surfaces. Keep in mind, however, once you have developed the skills taught in this chapter on how to modify a surface, particularly with polyurethane (PU), and combine it with a flow-through scaffold, you can imagine benefits in both environmental practice and medical applications (therapeutic and diagnostic). To indicate the scope of modifications, we will learn how to change a hydrophobic surface to hydrophilic. We will show research on making PU suitable for direct blood contact. Still further we will describe a way to immobilize enzymes to a scaffold in a one-step continuous process. This technique opens up economic and commercial opportunities in both medical and environmental uses.

Before we begin, we need to refine our definition of a scaffold. The chemical requirements of an ideal scaffold are for the most part independent of architectures, as many of the citations will illustrate. We will cover many of these without regard to the scaffold, but we don't want to drift too far from the goal of making a flow-through device with a compatible surface. Therefore, as we discuss the chemistries of scaffolds developed by researchers, we will include an evaluation of the scaffold with regard to the lessons taught in the last chapter. Chemical modification requirements of an ideal scaffold include a number of factors:

- *Chemical/biological stability*: Whether the application is environmental or medical, the conditions in which the device will be used can be harsh. Hydrolytic degradation of the scaffold and modifications thereof need attention. This is especially true for implantable devices. Toxic effects of degradation products are likely. Polymerized bioacids are commonly considered as "biodegradable" scaffolds, but the rate and products of the degradation have limited its development, not to mention the quality of the scaffolds. There are rigid requirements set by the FDA among other agencies that address the fate of any material that is intended for implantation. These requirements include extractables and cytotoxicity, both of which are directly related to the scaffold material. Particularly appropriate to this chapter, we will be describing the use of corrosive chemical to covalently bind active molecules to the surfaces. The chemistry itself might be effective, but it may destroy the scaffold.

- *Biocompatibility*: The issue of biocompatibility could take up a chapter in itself. We will cover it more completely on an *ad hoc* basis at some point, but for now we want to take a broad view of the subject. This view might be summarized by the philosophy that one "should do no harm."
- *Surface chemistry*: While we don't think the issue is settled, it is generally accepted that the surface be hydrophilic. We will discuss at length an exception that uses a hydrophobic PU on which hepatic cells in a matrix are adsorbed. Having said that, hydrophilic surfaces are considered "more" compatible. For the designer of a scaffold however, it is also a general rule that hydrophilic scaffolds are physically weak, while hydrophobic scaffolds are strong. In any case it reinforces the concept of separating the function of a successful scaffold device into a scaffold with physical strength and a hydrophilic surface treatment. In lectures we would joke that God approach the problem of designing humans by settling on a calcium-based scaffold supporting a variety of hydrogels.
- *Environmental*: Lastly, the material must function in the environment in which it is placed. Thus two characteristics must be considered: the effect of the environment on the material and the effect of the material on the environment.
- *Economics*: While not as important to medical technologies, the cost of materials can be an important factor in environmental processes. When the scaffold is composed of rocks or sand, it is less an issue, but when synthetic materials are used, including PU, the initial cost and durability are significant issues.

While we have digressed somewhat, we feel it is important to the theme of this book, we must keep in mind that this is not a book about scaffolds or immobilization. It is a book about the appropriate synergy of both.

Methods of Immobilization

A great number of materials and methods have been studied in the search for an ideal engineering scaffold. The accepted goal of these studies is that the scaffold for cell propagation should mimic, as much as possible, the native extracellular matrix (ECM). It must combine a functional surface, biocompatibility, and physical structure. Native ECM does more than providing a physical support for the cells. It provides adhesion proteins and regulates cellular growth. The current catalog of potential synthetic scaffolds is unlikely to induce cell adhesion and tissue formation by itself. To overcome this, the surface of the scaffolds must be modified to draw them closer naturally to native ECM. In this chapter we will discuss how the surfaces of synthetic scaffold materials can be modified to approach that goal. The techniques are summarized by the term immobilization. Several techniques are discussed.

It is useful to note that while the general category of immobilization is appropriate for both our medical and environmental readers, the specifics of the individual procedures may be more appropriate to one field or the other. To be sure the science is not directed to one group or the other. All scientists should be interested even if outside their particular focus. What differentiates them, as you will find, has more to do with the cost than the science. Specifically, when we discuss medical applications, the cost of many of the techniques would be prohibitive to almost any environmental use. Again it will be interesting to anyone but more academically rather than practically. Therefore when we get to examples of immobilizations, we will separate them into medical or environmental. First however is a general discussion of immobilization.

There are a number of ways in which the surface of a scaffold can be influenced. We will list and use examples to illustrate the techniques. Many if not all are appropriate for flow-through applications, but the examples are independent of that primary focus.

We need to mention that the process of immobilization affects the activity of the object that we are immobilizing. By way of example, by covalently immobilizing an enzyme, the useful life (expressed as the time it takes to reduce the activity by one half) is increased, but the activity itself is decreased. This is typically the case. In contemplating immobilization, one must be prepared to optimize the competing effects to yield an economic advantage. Table 4.1 shows examples of these competing influences. Again, the techniques are explained and illustrated following the table [2].

While each method you might anticipate requires a similar study, these data are typical. Let us now examine several techniques.

Table 4.1 Types of immobilization.

Technique	"Scaffold"	Activity (%)	Half-life (h)
Free cells	(None)	100	36
Entrapment	Alginate	54	8500
Entrapment	Acrylamide gel	22	570
Adsorption	Cellulose	97	400
Cross-linked	—	26	40
Entrapment	Carrageenan gel	44	38
Entrapment	Agar	57	27
Adsorption	Bone char	2	25

Source: Divinny *et al.* [1]. Reproduced with permission of Taylor & Francis.

Immobilization by Adsorption

The process of changing the surface of a scaffold can take a number of forms. The easiest method is by adsorption. This method involves the association of the biological with the surface (Figure 4.1). The association may be electrostatic, hydrogen bonding, or a combination of several factors. The bond may be very weak but sufficiently strong that it is used as long as the stresses of passing a fluid through the system are taken into account.

Bacterial cells, particularly, have developed surfaces that are designed to attach themselves to any hydrophilic or other surface. In an application that we will discuss shortly, the sludge from a municipal waste treatment plant was mixed with a reticulated polyurethane foam (PUF). Sufficient adhesion without any additional processing was achieved to make an efficient biotrickling filter. We also learned that increases in new sticky biomass caused the foam to become overloaded and collapse.

In these experiments the bond was sufficiently strong to permit high flow rates. A more common situation is that the bond limits the flow conditions. The flow rate, agitation, particle–particle abrasion, and other effects all have the potential to disturb the scaffold-absorbent bond. Temperature and in some cases pH must be considered.

Biofiltration

To illustrate the use of adsorption to achieve the desired changes in the surface of a scaffold, the following example can be considered typical. The most common use of the adsorption phenomenon is in environmental remediation, and it is this application that we will examine first. It serves as a model for adsorption in general and so it is a useful example. We have discussed in an earlier chapter the development of a PU composite. It is a hydrophilic PU coating on a reticulated PU substratum. The purpose as we discussed was to impart a hydrophilic surface on an otherwise hydrophobic base. In the chapter on scaffold, we made the case that the reticulated substratum had architectural

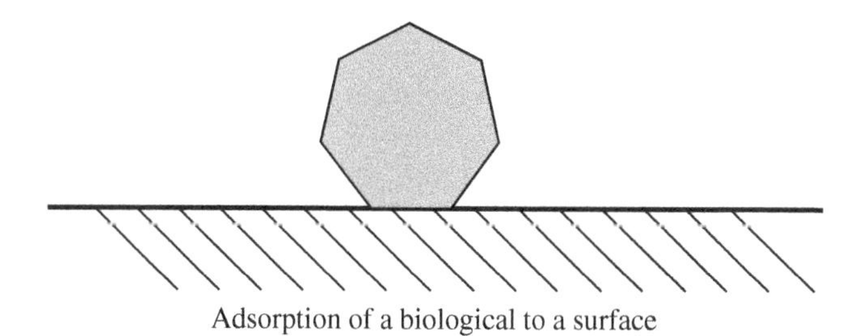

Figure 4.1 Adsorption to a surface.

properties that made it a candidate for flow-through applications. Coupling this with a hydrophilic surface would appear to have ideal properties for the following environmental application. We were interested in finding whether this composite foam could be used as a packing in what is called a biotrickling filter. The technique will be explained shortly. To investigate this new material, we sought the help of Dr Marc A. Deshusses, assistant professor of the Chemical and Environmental Engineering at the University of California (UC), Riverside. He is a recognized expert in the field with numerous papers and books [1]. Much of what we report here is based on his experiments as described in his report to us. Again the purpose of the work was to evaluate the overall applicability of the composite foam as a packing for biotrickling filters compared with conventional reticulated foam. The research conducted at Dr Deshusses' lab used toluene and H_2S as model pollutants. In addition, the performance of the foams was compared with other scaffolding materials.

The foams used for the initial experiments were characterized in our lab and the results were typical of our experience in these materials. The reticulated foam substratum was obtained from Rogers Foam (Somerville, MA, USA). It was used as the control and as the substratum for the hydrophilic coating. The coating technique is described in US Patent 6,617,014. Due to problems during the studies, a third sample was prepared in which the substratum was coated with an SBR latex, dried and then coated with the hydrophilic PU. This gave it more compressive strength, but it was clear that additional physical support was needed. That work was completed and tested with good success. Those data are not reported here, however.

Returning to the materials, the substratum and control was a conventional 20 pore per inch (ppi) reticulated for manufactured by FXI Corp. (Media, PA) and converted to sheets by Rogers Foam Corp., Summerville, MA, USA. The hydrophilic prepolymer we used was Hypol® 2002 and was obtained from Dow Chemical, Midland, MI, USA. The coating weight of the hydrophilic coating was 25.1%. It is calculated as the weight of hydrophilic coating divided by the weight of the composite. The equilibrium moisture was 37% and is calculated as the weight of the absorbed water divided by the weight of the wet composite. This latter measure is useful in evaluating hydrophilic materials and the technique is discussed in the chapter on laboratory practices.

The hydrophilic coating was done at our facility. An emulsion of the Hypol 2002 prepolymer was made from one part prepolymer and 2 parts of a 0.5% solution of Pluronic L-62 obtained from BASF Corp., USA. The emulsion was made in a pin mixer of our design and immediately deposited onto sheets of reticulated foam (0.7 cm thick and 45 cm wide, nominal). The equipment used for the coating was of our design and performed at our plant in Saco, ME, USA. It is shown here (Figure 4.2).

The mix head moves left and right, while the roller pulls the foam and presses the emulsion into the substratum. Sufficient curing takes about 10 min at which

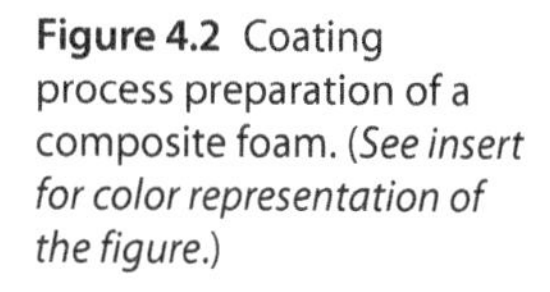

Figure 4.2 Coating process preparation of a composite foam. (*See insert for color representation of the figure.*)

point the material can be handled, but not rolled tightly as ultimate curing takes about 24 h. Drying is accomplished by passing warm air through the foam. The material was then rolled and packaged for shipment.

Biotrickling Filter Setup and Operating Conditions

In the first phase, four laboratory-scale biotrickling filters were constructed (again, at the UC, Riverside). Two were used for the treatment of H_2S (control and composite) and two were used for toluene treatment. The biotrickling filters were made of clear polyvinyl chloride (PVC) pipe (15.4 cm (6″) ID and total bed height of 121.9 cm). Each biotrickling filter was packed with foam. The result was a bed volume of 23.6 l. Schematics of the experimental setup are shown in Figure 4.3.

Table 4.2 presents the operating procedures for the hydrogen sulfide experiments and the initial toluene tests.

Prior to start-up the pressure drop versus airflow characteristics of one of the columns packed with modified foam was determined. This was to ensure that the packing was not compressed to the extent that its pores would be closed and result in excessive pressure drops. The pressure drop was very low indicating good airflow (Figure 4.4).

Colonizing a biofilter is called inoculation. In this study various sources of microorganisms were used. These included a soil extract, microorganism taken from other toluene-degrading biotrickling filters, and microorganisms from the field biotrickling filter operated at Orange County Sanitation District. Note that the source of the inoculum for toluene- or H_2S-degrading biotrickling filters is not critical. Toluene- and H_2S-degrading organisms are widespread and treatment performance is usually not affected by the nature of the inoculum, provided that it is diverse enough.

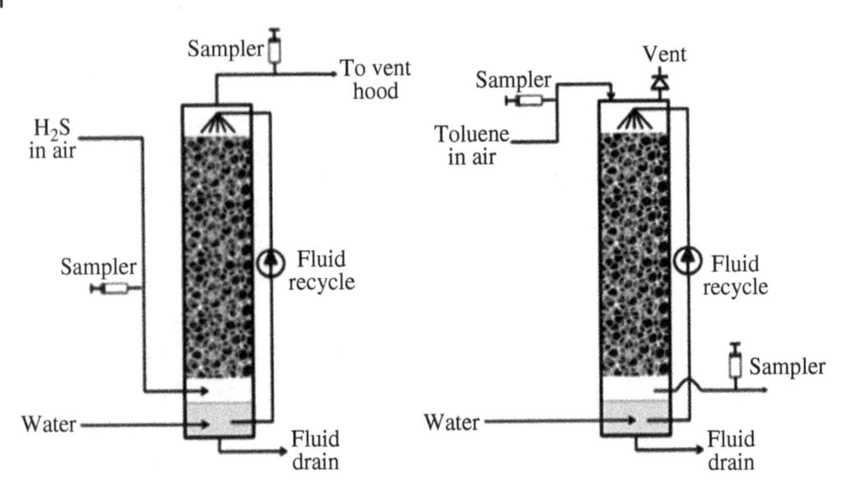

Figure 4.3 Biofilter setup for toluene (left) and hydrogen sulfide.

Table 4.2 Biofilter characteristics at start-up (unpublished data).

	Toluene reactor		**H_2S**	
Characteristic	**Control**	**Composite**	**Control**	**Composite**
Column height (m)	1.21		1.21	
Column diameter (m)	0.157		1.57	
Foam volume (M^3)	0.0236		0.0236	
Foam mass (g)	558	763	619.9	815.4
Foam density (kg/m^3)	23.7	32.3	26.3	34.5
Flow direction	Cocurrent		Countercurrent	
Flow rate (m^3/h)	1.4–2.8		1.4–2.0	
EBRT (s)	30		30	
Initial pollutant conc. (g·m^3)	0.4–1		200 ppm	
pH	8-Jun		1.6–2.0	
Operating temp (°C)	20			

The Toluene Reactor

After a short period of operation, the foams began to collapse. The reason is clear. Toluene is a carbon source for the bacteria and as a result the bacteria rapidly build biomass. The result was a decrease in porosity and pressure buildup, causing the premature termination of the study.

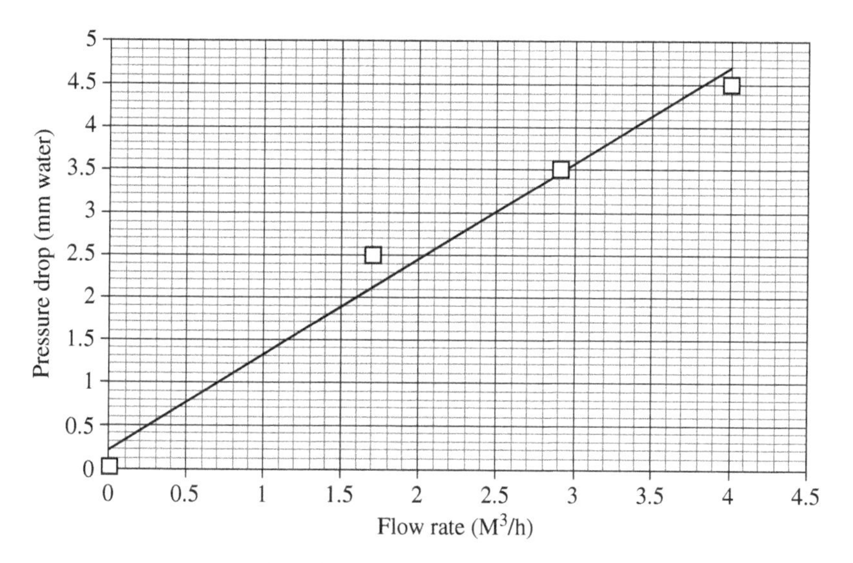

Figure 4.4 Pressure drop through composite foam.

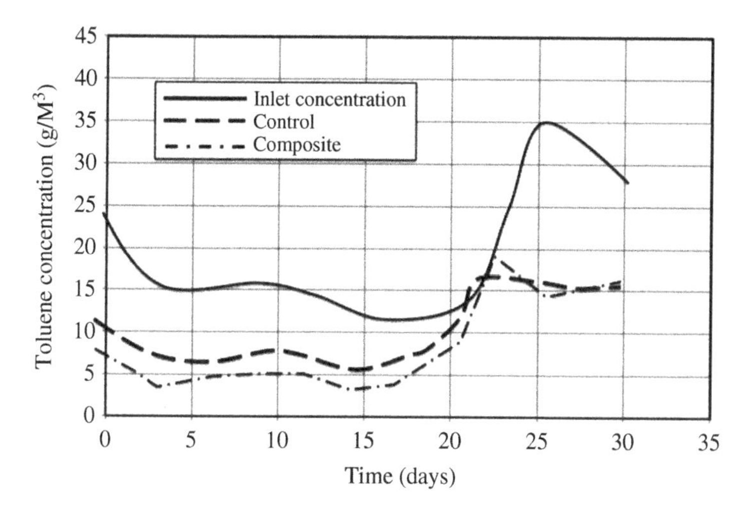

Figure 4.5 Start-up of toluene bioreactor.

Despite this, the start-up of the toluene-degrading biotrickling filters is shown in Figure 4.5. The graph shows that both biotrickling filters had a rapid start-up with effective toluene removal. This is most likely because both biotrickling filters were inoculated with a large density of microorganisms that were capable of degrading toluene. Examination of the figure seems to suggest that the reactor packed with the modified foam had better performance for the first 16 days. This would appear to support the conclusion that hydrophilic surfaces are preferred. However, after the first signs of the bed collapse (day 16), toluene removal decreased slightly and was comparable with the removal in the regular foam.

Recognizing the seriousness of the strength problem, we worked in our lab to mitigate it. As a result we developed a technique in which the reticulated substratum was pre-coated with a styrene–butadiene latex. While the latex we used is proprietary, in fact, any high glass transition grade will do. The treatment made the foam stiff enough to complete the experiments. We continued our strength studies in our lab, while the biofilter experiments were continued. Flow studies conducted in our lab confirmed significant increases in compressive strength, although no biofilter tests were done on these materials. It is important to note that while the pretreatment to stiffen the foam was critical to continuing the study, that surface that the organisms "sees" is the hydrophilic PU.

The toluene experiments were restarted with the stiffer foam. It was confirmed that the new foam packing was stiff enough to continue, but the trickling rate was reduced to decrease the actual water held in the packing. The strength allowed for sufficient time for determining the performance under pseudo-steady state. The bed, however, collapsed after 56–65 days of operation.

We decided that rather than using the control foam in the initial studies, it would be more commercially significant to use a PUF that is currently used in trickling biofilters. A foam from Zander (Germany) was selected. The foams are reticulated with larger pores and are significantly stiffer than the test foams. We will discuss this material more completely later in relation to a full-scale odor reduction system. For this study, however, the 4 cm cubes of foam were loaded into the control column in a random dump manner. Compare this method with the stiff foam that was rolled from a continuous sheet and inserted into the column. Liquid flowing to the stiff foam was forced to go through the foam, while the Zander foam had flow patterns around the cubes. This was mitigated to some degree by larger pore size. We will address this effect when we discuss studies at the University of Maine on fish tank water treatment. Also the smaller pore size of the composite foam would imply a higher surface area. These factors meant that the relative efficacy of the two foams cannot be determined. Despite these complications, the experiments were instructive with regard to the efficacy of PU and perhaps the advantage of a hydrophilic surface. This latter property is more clearly demonstrated in the H_2S studies.

Before we get to that, an examination of the results of the toluene degradation studies is necessary. There are many ways to show efficacy, but the most convenient was to show the outlet concentration as a function of the inlet concentration. These data are shown in Figure 4.6. Data was gathered on all

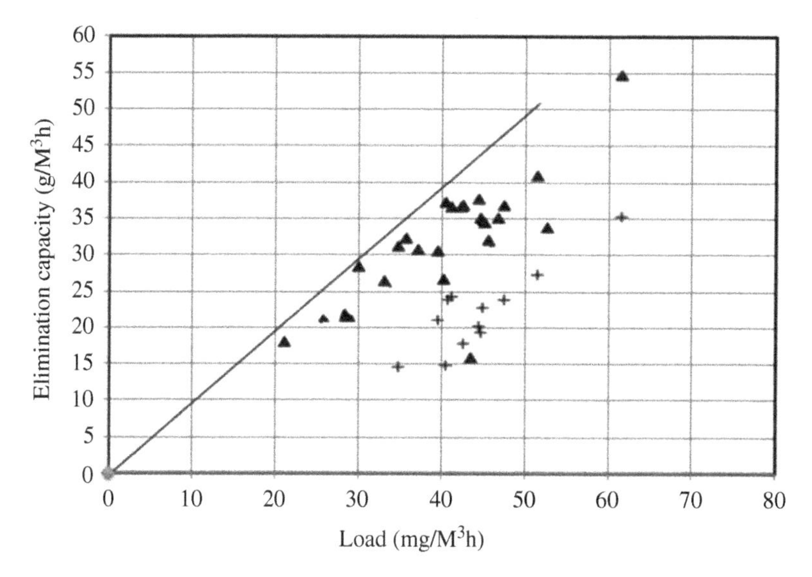

Figure 4.6 Comparison of the performance of the modified composite foam (▲) and the Zander (+) foam.

the foams, but only the stiff foam established some level of equilibrium sufficiently long to give good confidence. While there is difference in the effectiveness of the stiffened foam compared with the Zander product, that conclusion would be incorrect. The two foams are too dissimilar to be compared that way. While the stiff foam is relatively hydrophilic, that is not demonstrated in this study. What is probably demonstrated are differences in column packing and net surface area. When we discuss the degradation of hydrogen sulfide, the issue of compatibility with hydrophilic surfaces is seen more clearly.

Toluene is probably the most studied compound in biotrickling filters. Table 4.3 lists several studies along with the results of the present work. It is taken from the confidential report without references. It is included here for comparison.

Not unexpectedly, the maximum elimination capacity varies greatly. This is due to variations in surface area, fluid velocity, type of inoculum, nutrients (other than the target compound), reactor design, and, of course, type of packing. In addition, the operation of biotrickling filters needs to be considered including gas residence time, inlet concentration, and cocurrent versus countercurrent operation. Despite that, the comparisons with the data in this report provide a market perspective for reticulated PU. The data obtained with composite foam falls in the average of the range of the performance of the other methods (confidential report).

Table 4.3 Various studies on the removal of toluene (information is part of an unpublished research).

Characteristics	Packing type	Critical load (g/m^3h)	Maximum elimination capacity (g/m^3h)
Biotrickling filter (this study)	Regular foam	<18	40 to >50
Biotrickling filter (this study)	Initial modified foam	<18	40 to >50
Biotrickling filter (this study)	Stiffer modified foam	<18	40 to >50
Biotrickling filter (this study)	Zander foam	<20	>40
Biofilter	Compost + wood chips	<30	Oct-40
Biofilter	Compost + polystyrene beads	<10	20–25
Biofilter	Compost + perlite	30–40	45–55
Biofilter	Compost + perlite	10-Jun	23–32
Biofilter	Compost	N/A	100
Biotrickling filter	Pall rings	<25	71–83
Biotrickling filter	Crushed Pall rings	<30	79 ± 6
Biotrickling filter	Pall rings	ND	35
Biotrickling filter	Celite pellets	ND	60

The H_2S Reactor

The hydrogen sulfide removal tests were in general more successful than those performed with toluene. The height of the foam beds remained constant for about 120 days, after which a slight decrease in bed height (2–5 inches) was observed. The reason for the difference between the toluene- and the H_2S-degrading biotrickling filters lies in the microbiology of the process culture. Toluene degraders are heterotrophic organisms, that is, they use toluene as their carbon and energy source. This results in an increase in biomass. On the other hand, H_2S use atmospheric CO_2 as their carbon source and they get energy from the oxidation of sulfide. The growth rate of biomass, therefore, is slow compared with toluene.

The H_2S removal tests were conducted as follows. The columns were packed with the original foams, that is, not the stiffened foam. First, the biotrickling filters were allowed to reach a steady state, while the experimental setup was tested and refined. After about 40 days of treatment, near complete removal of H_2S was observed, and the setup was modified in order to be able to test higher

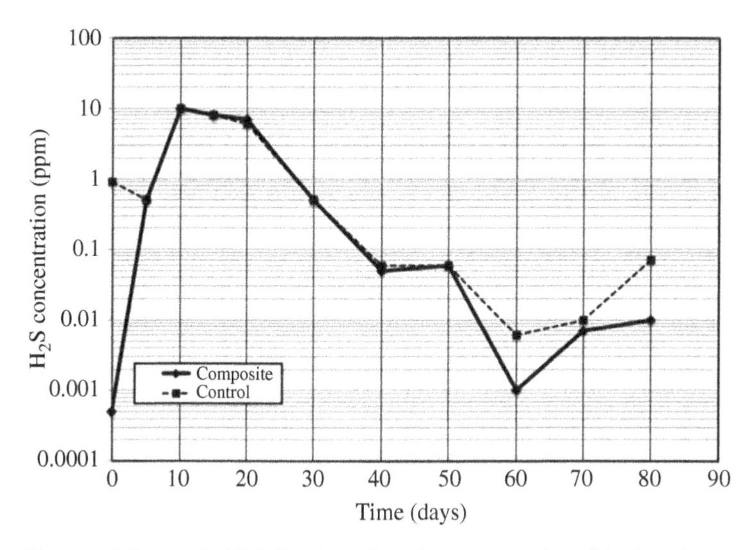

Figure 4.7 Removal of H_2S. (*See insert for color representation of the figure.*)

flow rates and higher H_2S inlet concentrations. Finally, a series of tests were conducted at four different airflows where the inlet H_2S concentration was varied and the outlet concentration was monitored. The experiment focused on determining the elimination of H_2S at high concentration in order to observe performance at high elimination capacity. The experiment was also aimed at finding the maximum inlet H_2S concentration that would result in an acceptable outlet concentration, which was set at about 1 ppm.

The biotrickling filter behavior during the start-up phase is shown in Figure 4.7. Interestingly, removal of H_2S was observed 24 h after starting the systems. Thereafter, minor problems in the H_2S system and low feeding of makeup water may have caused some instability in the removal, but when these were corrected, stable and effective removal of H_2S occurred. Inlet and outlet concentration data are shown on a log scale. The control and composite foams behaved nearly identically until about day 50. After that point, the reactor with the composite foam exhibited lower outlet concentration. The difference was considered significant.

The filters were tested at several different airflows and various concentrations. The data plots the inlet concentration against the outlet concentration as a function of the empty bed residence time. This is essentially a study of reaction time.

While not evident on this scale, Figure 4.8 shows the reactor packed with the coated foam that is lower than the outlet of the regular foam.

In summary, the elimination capacity versus load curve is shown. The graphs show that removal was essentially complete at all conditions tested (Figure 4.9).

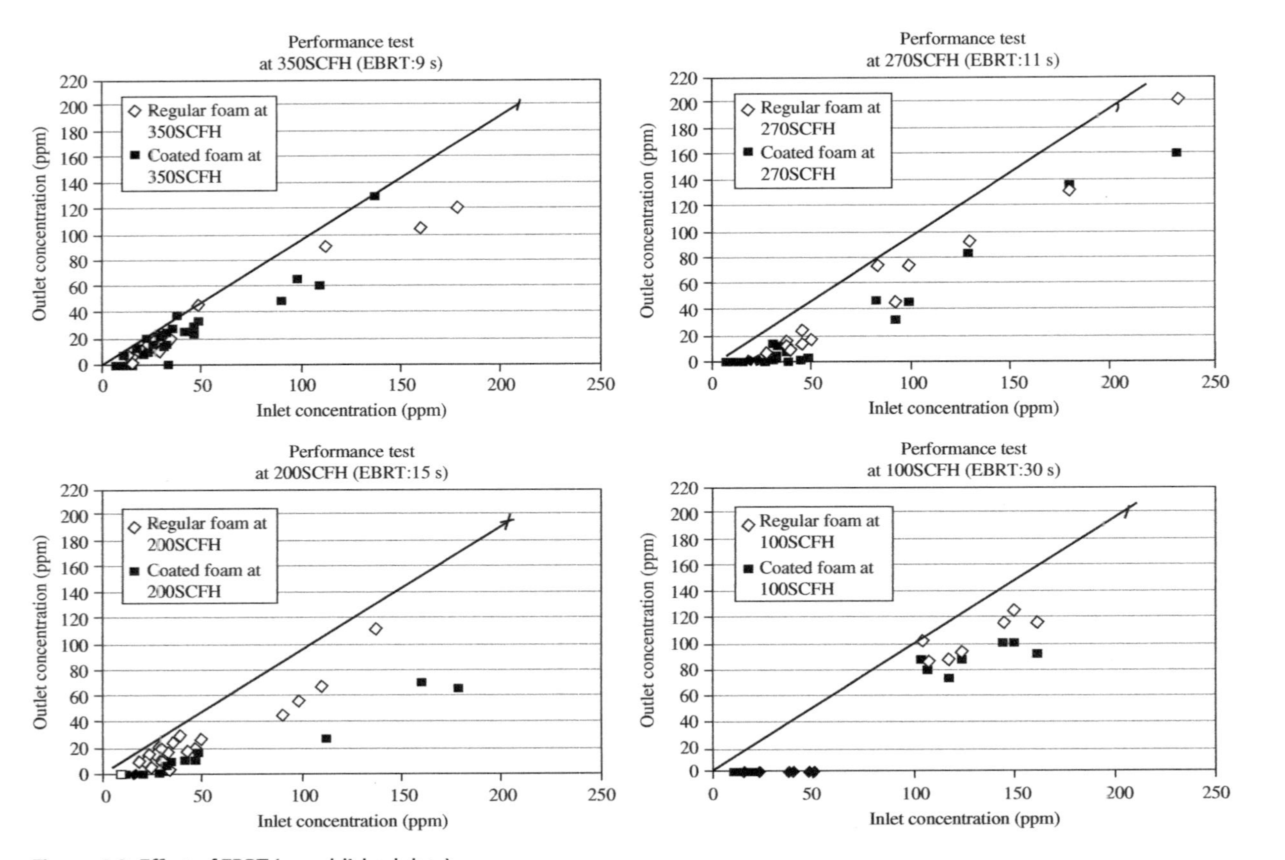

Figure 4.8 Effect of EBRT (unpublished data).

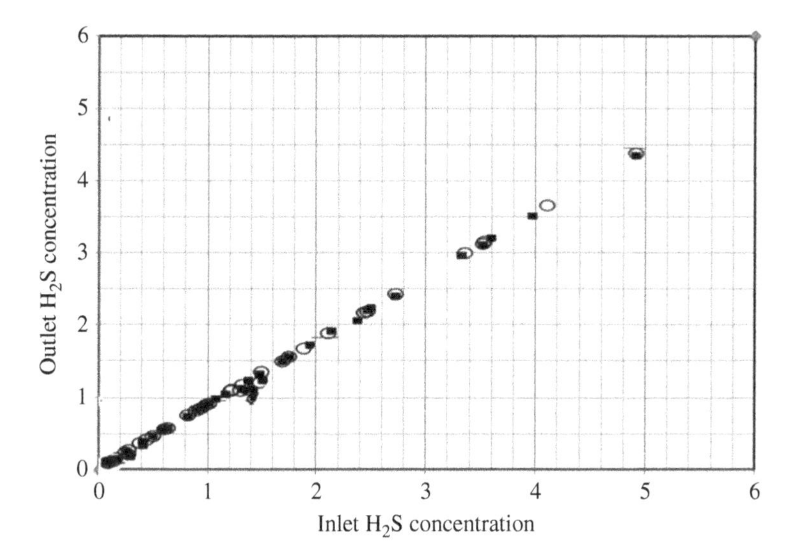

Figure 4.9 Performance of the H_2S bioreactor.

Summarizing the UC study, the biotrickling filter packed with the PUFs performed reasonably well. Under several operating conditions, excellent removal (>99%) of H_2S was obtained. Thus the foams were considered a suitable packing material with certain reservations. Inasmuch as this was the first time these foams were used in a biotrickling filter, one might suspect that characteristics discussed earlier would improve the efficacy. We will be discussing other immobilization techniques that might enhance the surface above and beyond hydrophilicity.

Experimental observations led to the conclusion that the biotrickling filter packed with the hydrophilic coating exhibited systematically better H_2S removal than the one packed with regular foam. It is reasonable to conclude that the modifications made to the foam were responsible for these differences in performance. The difference in H_2S removal was small, but could have significant implication as far as reactor sizing is concerned.

Compared with other available scaffolds (foam, lava rock, structured packing, random dump), the composite foam appears to be a good support, outperforming other scaffolds in some, but not all, cases.

Biological Treatment of Aquarium Tanks

The University of Maine conducted a series of trials on the effectiveness of a carbon/hydrophilic PU filter medium for recirculating aquaculture tanks [3]. A hydrophilic prepolymer was supplied by Rynel Corp. of Boothbay, ME, USA.

While not precisely the same, it is safe to assume that the prepolymer was equivalent to Hypol 2000. The foams were produced by mixing activated carbon, water and the prepolymer. The material was used as a biofilter medium in recirculating fish tanks by providing a surface for the attachment, growth, and multiplication of nitrifying bacteria. Water treatment is necessary to remove inorganic nitrogenous compounds from the water, particularly ammonia and nitrite, both of which can be toxic to the cultured species.

Atlantic salmon was the species chosen for these experiments based on previous experience with this species. Two recirculating systems were designed. One tank was used for this study and the other as a control.

The fish tank drained into a settling tank. The water in the settling tank was pumped into the trickling biofilter that flowed into a head tank. The head tank maintained a constant flow to the fish tank. A UV sterilizer (not shown) was placed in the line between the head tank and the fish tank (Figure 4.10). The biofilter foam was cut into cubes of approximately 50 mm^2.

At week 1, each system was stocked with five fish averaging 530 g each. At week 3, five more fish were added to each system, and this procedure was repeated every week up to the seventh week to give a total of 30 fish per system. Feed level during this time was kept constant at 0.5% bodyweight. The feed was metered out by hand so that feeding behavior could be observed and to make sure that all the feed was consumed. At week 8 the feed rates were increased until ammonia levels reached 1.0 mg/l. At this point the feed level was kept constant for 2 days and then reduced by half for the remaining weeks of the trial.

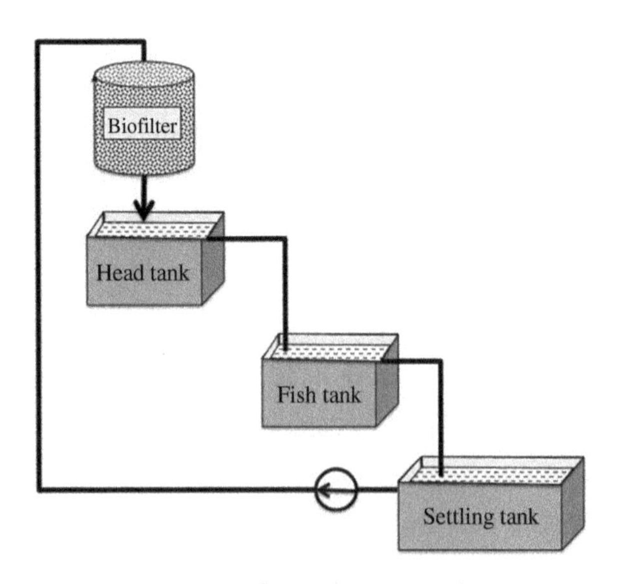

Figure 4.10 University of Maine Aquarium study.

The ammonia and nitrite showed gradually increasing concentrations up to a maximum of approximately 1.0 mg/l. At this point the biomass density was 73 kg fish/m^3. The feed rate was approximately 0.5% bodyweight. The mean maximum weekly ammonia level was 0.782 mg/l although some individual readings reached 1.2 mg/l. Mean maximum nitrite concentration was 1.27 mg/l with some readings over 1.5 mg/l.

According to the authors, the experimental material performed well as a biofilter medium with the additional benefits of providing suspended solids reduction and removal of dissolved organic compounds. Not included in the study was a control using only hydrophilic foam. As we know, hydrophilic foam is by itself capable of supporting a bacterial colony. What was mentioned is an important part of the permeable scaffold concept. They discussed the surface area of the individual pieces of foam. An electron microscopy examination showed the absence of nitrifying bacteria within the foam. In our opinion this is primarily due to the lack of flow-through. It is clear that most of the flow passed over the surfaces of the foam pieces. Despite this inefficiency, the foam performed well.

Protein Adsorption

We will now shift our attention away from the adsorption of bacterial cells and the example of how that is used in the remediation of contaminated water to a more general study of the adsorption of biomolecules (specifically proteins) to a surface. While considered a problem in most cases, the adsorption of proteins on an otherwise uncontaminated surface is unavoidable. Those of you who use contact lenses are familiar with the phenomenon. Boat owners are aware of the fouling by subsurface organisms and slimes. The food processing industry must be continuously vigilant in control of the problem, not only of residual food but also of bacterial growth.

When a biological fluid comes in contact with almost any surface, proteins and similar biomolecules migrate from the fluid to adhere to the surface. The bond that is created might be strong enough to withstand normal external stresses, but in some cases the bond is nearly irreversible. In either case, the effect is a change in the surface chemistry. In the case of adsorption to a membrane, the permeability is effected. When the biological fluid is blood, a multitude of effects are possible. This includes inflammation and or encapsulation.

While these examples are mostly problematic, there are examples that are critically important to the protection of our environment. Specific is in the remediation of air and water. In this case we take advantage of the natural ability of bacteria to adhere to almost any surface. Before we get to that, while it is the theme of this book to teach how one immobilizes biologicals, for instance, sometimes it is important to prevent adsorption. Preparing a surface for a controlled immobilization involves cleaning it. In that case a procedure to inhibit protein adsorption might be appropriate. An example just mentioned deserves elaboration. For implantable devices, uncontrolled contamination of a surface by proteins can contribute to the activation of platelets, which can lead to inflammation, thrombus formation, and encapsulation. For this reason we will devote a few paragraphs on preventing immobilization. If we are to immobilize, we certainly need to control what gets immobilized.

Among the methods to inhibit protein adsorption is a pretreatment with, for instance, polyethylene oxide (PEO) and analogs thereof. Many theories are presented as to why this is effective, including the insulation of the proteins from the influence of the otherwise hydrophilic surface [4].

Several methods have been suggested in order to achieve a PEO surface. Presented here is a method developed by Hubble *et al.* to affect an interpenetrating network of both PEO and polyethylene terephthalate (PET). The key to the technique is a solvent for both the scaffold and the PEO or PET. Acetone is an example that is useful for many hydrophilic surfaces. For a polyether-based hydrophobic PU, tetrahydrofuran (THF) is used. In either case, the scaffold is dissolved and a film is cast. The resultant film is immersed in the PEO or PET solution to affect interpenetrating adsorption. We, of course, will be looking at a PUF and this procedure is appropriate. Our goal is to combine chemistry with architecture, and thus dissolving the PU destroys the former. A less destructive method could be developed, but we will show that this may not be necessary.

One of the most remarkable examples of immobilization by adsorption is the combination of the protein avidin and its affinity for the vitamin biotin (also known as vitamin B_7). We had mentioned that some of the examples would be outside the usefulness of environmental remediation for cost reasons. This is the example we were thinking of. The chemicals used to affect these procedures are sold in gram quantities. Nevertheless, the technique is interesting even to us industrial chemists.

The Avidin–Biotin System

The full scope of this system is beyond the focus of this book, but it would not be appropriate to not cover it in sufficient detail to show its usefulness.

Avidin is a protein derived from the eggs of avians, amphibians, and reptiles [5]. It has become an important tool in the investigation of many biochemical processes. Its function is based on its affinity for biotin. Biotin is also known by its more familiar name, vitamin B7. As we will show, the vitamin can be derivatized to perform a particular function. An example might be the separation of an antibody by derivatizing the biotin with the complement antigen. Avidin and similar proteins (streptavidin, for instance) have the ability to bind up to four biotin molecules. It is this ability to bind (with remarkable strength) derivatized biotins that makes it ideal for both purification and analysis of biological systems.

Avidin was first discovered by the observation that chicks on a diet of raw egg white were deficient in the vitamin. It was concluded that a component of the egg white was responsible for biotin binding. Later it was confirmed that the deficiency (at the time was known as "egg white injury") was caused by the protein avidin. For clarity we will use the most commonly used term biotin and its derivatives to represent the vitamin component of the avidin system.

While the application of this technique has taken many paths, for our purpose, we will focus on the immobilization aspect and specifically biologically active molecules. As we will show, the process by which the immobilization begins is the association of the avidin protein to the scaffold we have designed. This association is adsorption as is the binding of the biotin to the avidin, keeping with the theme of this section. It must be added however that the avidin–biotin system also depends on the covalent modification of the biotin.

For our purposes the avidin will be described as a pretreatment for the scaffold. This will be clear from the examples we will present. Biotin and its derivatives will function as the active part of the system. That is, it will be the reason we are designing a functioning scaffold. Put another way, the avidin binds to both the scaffold and the biotin by adsorption, while the biotin binds biologically active molecules such as an antibody or enzyme. The list of biologically active molecules includes but is not limited to mammalian and bacterial cells, yeasts, molds, hormones, nucleic acid drugs, and toxins. Lastly, and returning to the discovery of this process, it was the egg white diet of the chicks and more specifically the avidin component of the egg whites that sequestered the biotin that cause the deficiency. In other words avidin was the process by which the activity of the vitamin was reduced. We will use this same process but with bioactive molecules and cells bound to the vitamin.

Biotin itself is shown in Figure 4.11.

For instance, a target molecule (an antibody, for instance) can be isolated from a solution by immobilizing a biotin derivative and mixing it with the antibody solution. The derivatized biotin such as biotinyl *N*-hydroxysuccinimide ester is used in the biotinylation of proteins and peptide in the pH range of 6.5–8.5. Cells can be separated and suspended in a phosphate buffer by treatment with the same derivatized biotin.

Figure 4.11 Biotin.

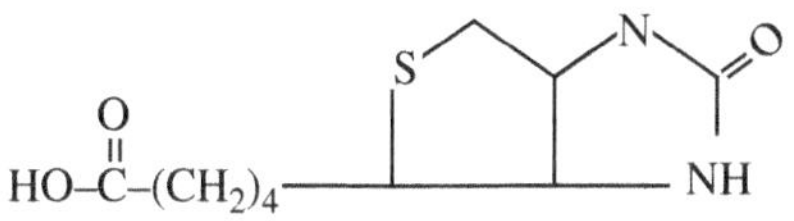

The sequestering of the target biological in the aforementioned examples is not the biological itself but rather segments of the biological. This is the power of the technique. Rather than attaching to the walls of the cells, for instance, the biotinylation, as it is known, is an association to amino acid residues on the surface. This explains why the biotinyl *N*-hydroxysuccinimide ester has found broad application. The biotinylation of proteins on cysteine residues is accomplished by using maleimidopropionyl biotin. The same process is used. Consider the following examples.

Application of the Avidin–Biotin System to Cell Adhesion to a Scaffold

Cell adhesion to a scaffold is a prerequisite for tissue engineering. Many studies have been focused on enhancing cell adhesion to synthetic materials that are used for scaffold fabrication. In the following study, the avidin–biotin binding system is applied to chondrocyte adhesion to several biodegradable polymers, specifically poly(D,L-lactic acid) and polycaprolactone [6].

The polymers were dissolved in dichloromethane. The solutions were then added to poly(propylene)-coated 96-well plates and the solvent evaporated. The surfaces were treated with a 0.05 M sodium hydroxide solution. The polymer-coated wells were sterilized with 70% ethanol and exposed to UV overnight. Cell adhesion studies later showed that there was no substantial effect on cell adhesion by this treatment.

Porcine chondrocytes were biotinylated by incubating the cells in a phosphate buffer with 3-sulfo-NHS-biotin (Sigma, Cat. no. B5161) for 30 min. An avidin solution in a phosphate buffer was added to each well. The wells were seeded with the biotinylated chondrocytes. The untreated polymers did not support chondrocyte adhesion well as shown in Figure 4.12 (restated from the original paper for clarity). The avidin-coated PDLLA and PC showed increased adhesion.

One must remember, however, that the avidin system does not sequester the cells but rather the amino acid residues on the surface of the cells.

It is appropriate that we discuss this chemistry outside the context of scaffold as we described it. While it is an excellent technique and has analytical possibilities, to be useful in the broader sense, the technique must be applied to more clinically appropriate scaffolding. It is as if a cake recipe describes the icing in great detail without describing to cake it is intended to cover. The following paper applies this immobilization to a true scaffold.

Adsorption to a Tricalcium Phosphate (TCP) Scaffold Using the Avidin–Biotin System

A notable exception (other than this book) to the lack of scaffold information is the following work by researchers at Zhongshan Hospital, Fudan University, Shanghai, China [7]. In this study, a scaffold is developed, characterized, and

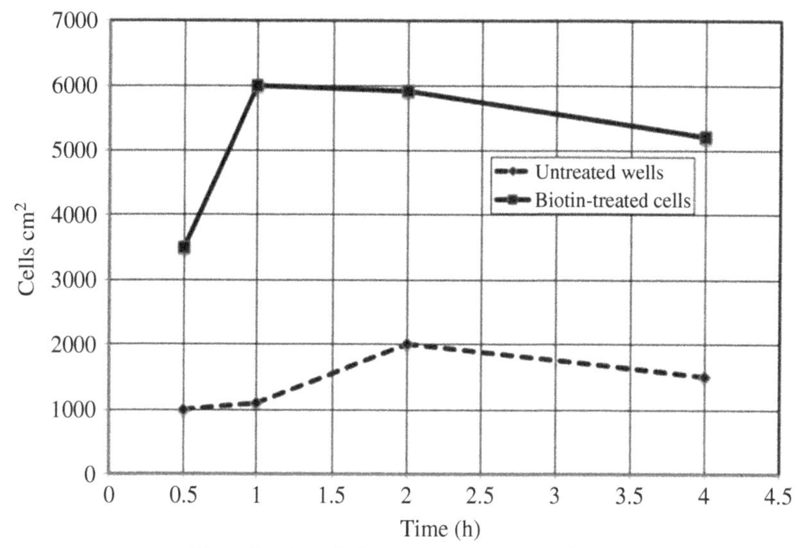

Figure 4.12 Adhesion of chondrocytes.

found to have many of the properties discussed as important to the task. This was followed by the adsorption of an avidin–biotin system. The composite is then evaluated. This will serve as a model when we discuss our work on PU composites.

In their study they investigated the properties of a scaffold for bone repair using adipose-derived stem cells (ADSC). The scaffold was of tricalcium phosphate (TCP) that was avidin-coated and reacted with biotinylated stem cells. The method of preparation of the scaffold will be described here as will the analysis to the physical nature of the architecture. While the immobilization to the scaffold is similar to that in the other examples, we will discuss their techniques. We hope to show that it is simple relative to the other immobilization techniques we will be discussing (e.g., covalent attachment).

The TCP scaffold was synthesized by the precipitation of calcium nitrate and diammonium hydrogen phosphate. The resultant precipitate was dried and calcinated in an alumina crucible at 700°C. The result was TCP crystals.

The TCP was mixed with two particle size grades of ammonium chloride (AC). This was to produce two distinct pore sizes within the scaffold. The blend of TCP and the ACs was placed in a mold and compressed. It was then sintered gradually to 1100°C.

While not to the analytical standards that will be required to advance the science of scaffold chemistry, the analysis of the resultant scaffold is sufficient to begin the comparison with other proposed constructs, including PUs. In this example, SEM micrographs "suggested" that the porous TCP scaffold had high pore interconnectivity and "abundant" pore structures, including the desired macropores (diameter, 200–300 μm) and micropores (size, 60–100 μm). The structure had a high surface area and the void volume approached 71%. The compressive strength was an impressive 7.9 MPa (Figure 4.13).

Preparation of the Scaffold

ADSC were isolated and separated into two portions. One portion was treated with 3-sulfo-NHS-biotin (Pierce Biotechnology, Inc, Rockford, IL, USA) in a CO_2 incubator for 30 min (referred to as Bio-TCP). The other portion was left untreated as a control.

A sample of the TCP was soaked in an avidinylation reagent (Pierce Biotechnology, Inc) at room temperature for 2 h (referred to as Avi-TCP).

Four scaffolds were treated with the ADSC according to Table 4.4.

Following seeding, the samples were incubated for 1, 12, and 24 h at 37°C in 100% humidity and 5% CO_2. After culturing, the adhesion of the cells to the scaffold was measured. For the complete analysis, refer to the paper. For our purpose, however, this serves as a useful example of the adsorption process. The following chart shows the findings and supports their conclusion (Figure 4.14).

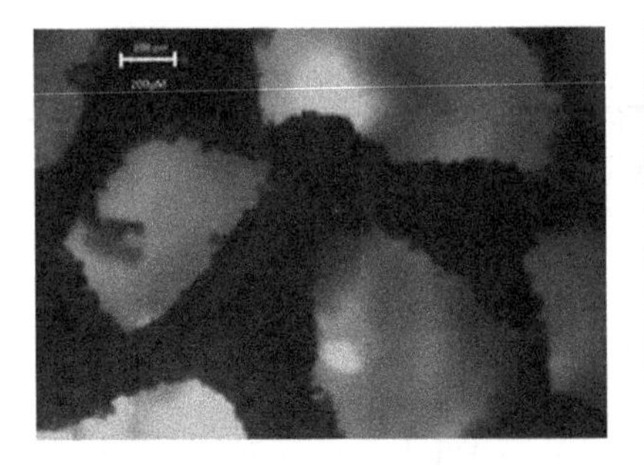

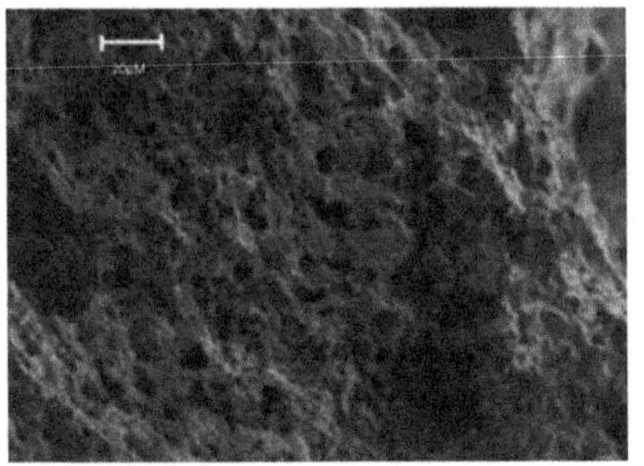

Figure 4.13 The macro- and microstructures of the TCP scaffolds. (*See insert for color representation of the figure.*)

Table 4.4 Schedule of samples for inoculation.

Cell	Scaffold
Untreated ADSC	Untreated scaffold
Untreated ADSC	Avi-TCP
Bio-ADSC	Untreated scaffold
Bio-ADSC	Avi-TCP

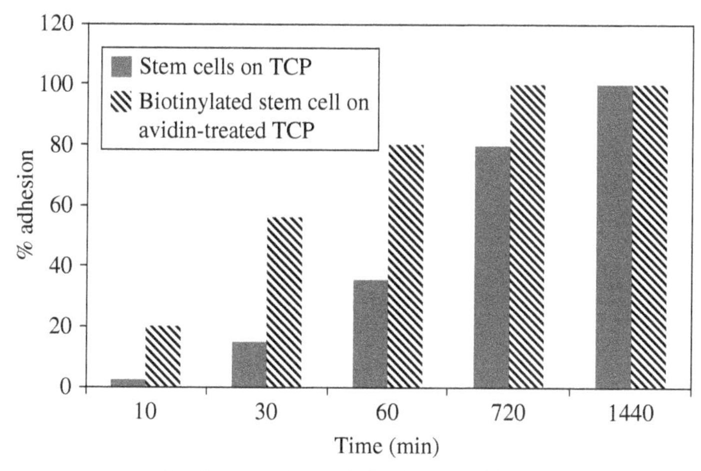

Figure 4.14 Results of cell growth. *Source:* Huang *et al.* [8]. Reproduced with permission of Elsevier.

Hepatic Cells on a Fabricated Polycaprolactone Scaffold

In another study, the avidin–biotin system was used to culture hepatic cells on a fabricated polycaprolactone scaffold. Selective laser sintering (SLS) was used to fuse the polymer (mixed with sodium chloride to create pores) [8]. SLS involves the use of a high power laser to fuse small particles of plastic into a mass that has a desired three-dimensional (3D) shape. The laser selectively fuses powdered material by scanning cross sections generated from a 3D digital representation of the part (e.g. from a CAD file). After each layer is scanned, the powder bed is lowered by one layer thickness and the process is repeated until the part is completed.

Hep G2 cells were seeded into the scaffold using the avidin–biotin adhesion method described in the last example. The importance of this study was another example of a research group considering the system, that is, the culturing of cells on a 3D scaffold with flow-through characteristics. In the present case, the void volume approached 90%. In another relevant application of adsorption, the attachment of ADSC to biomaterials is investigated for possible strategies for mediating inflammation and wound healing. The ADSC percent coverage was measured on common medical-grade materials subjected to different surface treatments. Cell coverage on silicone elastomer (polydimethylsiloxane) was below 20% for all surface treatments. Polyimide (Kapton), PU (Pellethane), and tissue culture polystyrene all exhibited more than 50%

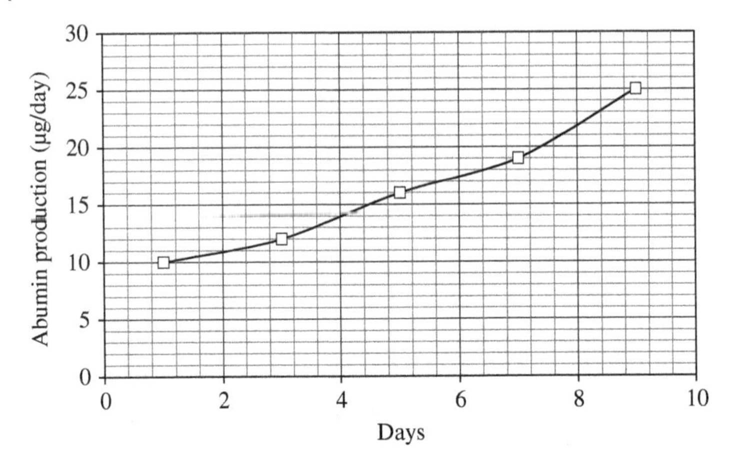

Figure 4.15 Albumin production of Hep G2 cells in the polycaprolactam scaffold. *Source:* Yamashita *et al.* [9]. Reproduced with kind permission of Dr. Yamashita.

coverage for surfaces treated with fibronectin, fibronectin plus avidin/biotin, and oxygen plasma plus fibronectin treatments. In culturing studies, after the first day, the fibronectin plus avidin–biotin attachment method gave significantly higher cell coverage than the control, while the fibronectin did not significantly increase the cell area coverage. However, by day 5, all of the attachment methods had increased and covered approximately 85% of the cell culture dish (Figure 4.15).

These data again show that the avidin–biotin system gives a faster start to the culturing process but other techniques eventually catch up. While not discussed we wonder if there is a concentration factor.

Summary of Immobilization by Adsorption

We could go on with the seemingly endless use of adsorption to affect immobilization, but the aforementioned examples show the technique both in environmental remediation and in the medical arena. We need to move onto other techniques, but we think it is useful to describe one more application. The reason it combines an overall purpose of the book is to juxtapose a PU scaffold with the immobilization of hepatic cell. We will make the case that a reticulated PU scaffold that is modified to have a hydrophilic surface is as close to an ideal synthetic scaffold. In the last example we will report on a research in Japan that recognized the benefits of the scaffold and used it to develop a functioning liver in large animal testing. We will describe the basics of their technique here as an example of the adsorption phenomenon but cover it in more detail later.

Figure 1.16 Coating process for the hydrophilic polyurethane composite.

Polyurethane Immobilization of Cells and Biomolecules: Medical and Environmental Applications, First Edition. T. Thomson.

Figure 4.2 Coating process preparation of a composite foam.

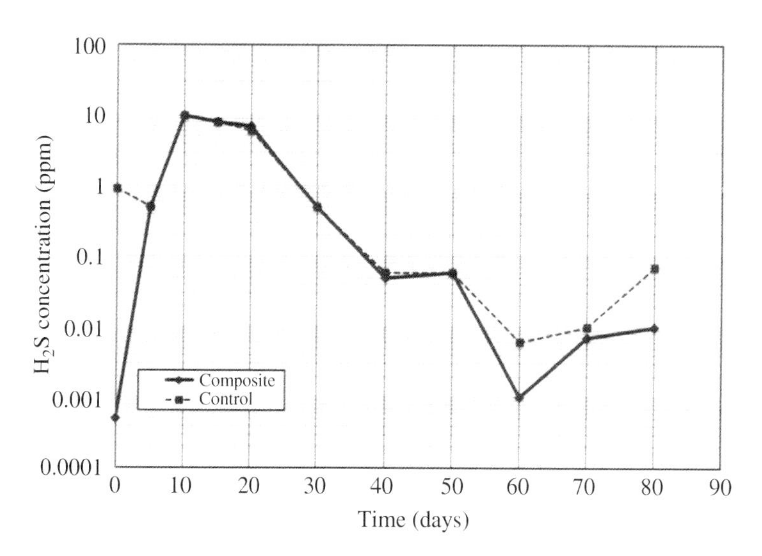

Figure 4.7 Removal of H_2S.

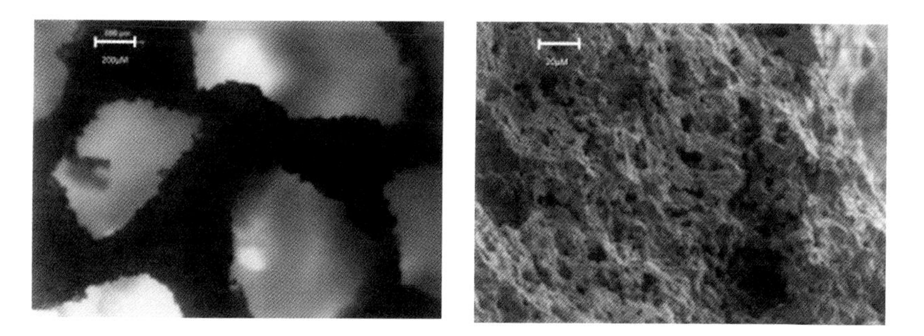

Figure 4.13 The macro- and microstructures of the TCP scaffolds.

Figure 4.18 Improved absorption of oil by a derivatized polyurethane foam (left).

Figure 4.22 Extraction of a dye using hydrophilic polyurethane.

Figure 4.23 Rotational spectroscopy using a fiber optics.

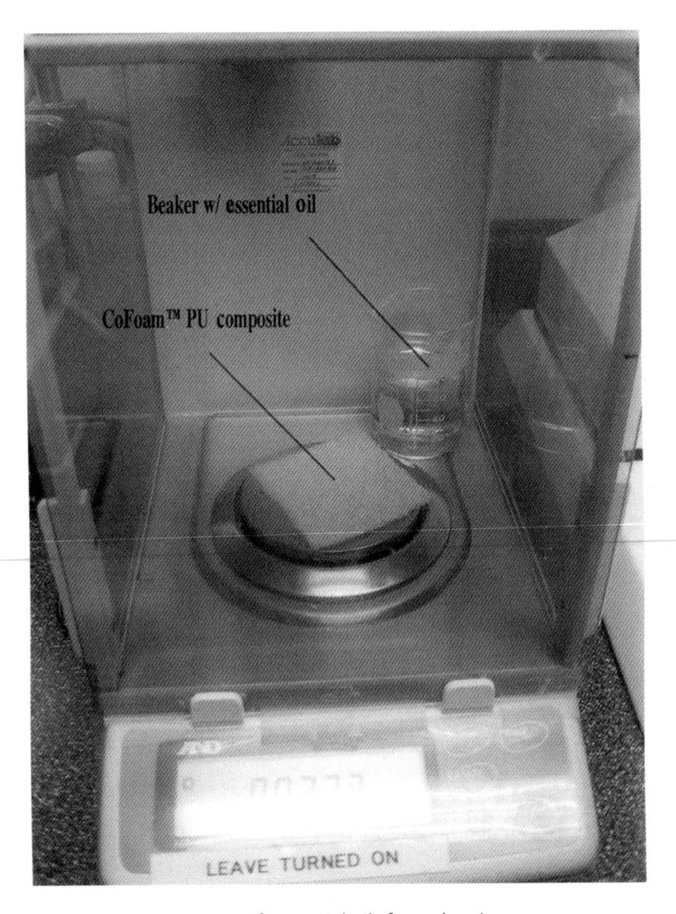

Figure 4.26 Extraction of essential oils from the air.

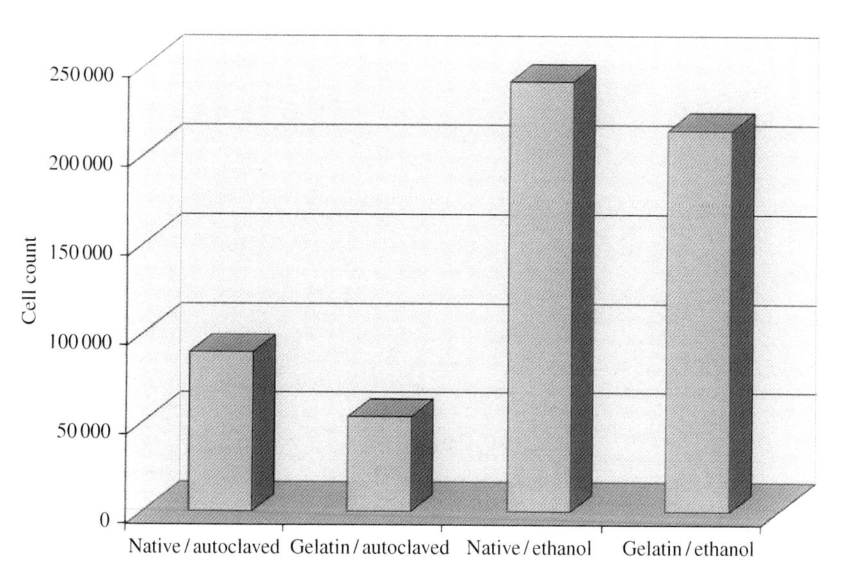

Figure 4.41 Cell growth of smooth muscles by composite scaffold by pretreatment.

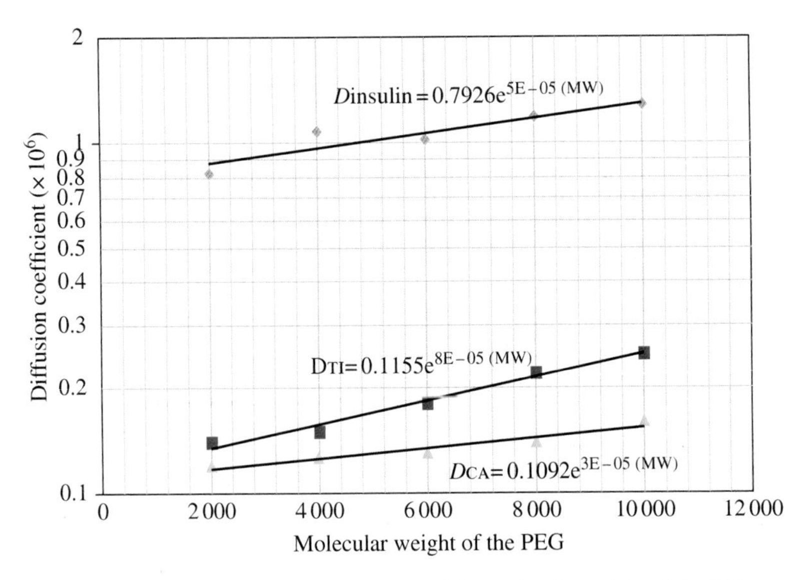

Figure 5.8 Diffusion coefficient as a function of the molecular weight of the PEG. *Source:* Data extracted from Weber et al. [5].

Researchers at Kyushu University in Japan placed primary human and primary porcine hepatocytes in a PUF/spheroid culture system and compared the metabolic functions in the plasma of patients with FHF in a 24 h stationary culture to those in a monolayer culture [9]. The PUF/spheroid culture system using primary human and primary porcine hepatocytes significantly decreased ammonia content during 28-day culture. The present results indicate that the PUF/spheroid culture system using primary human or primary porcine hepatocytes demonstrated more advantageous metabolic functions in the plasma from patients with fulminant hepatic failure (FHF) than the monolayer culture. The foam had an average pore size about 300 m, with a void volume of 98.8% [10].

In subsequent studies, hepatocytes of various animals such as rat, dog, and pig that were inoculated into PUF spontaneously formed spheroids with a range of 100–150 m in diameter within 24 h of culture. The spheroids formed maintained liver-specific functions such as albumin secretion, urea synthesis, and drug metabolism for at least 2 weeks. These spheroids can be used as a biocatalyst because they are immobilized on the internal surfaces of PUF pores.

Immobilization by Extraction

While not thermodynamically different than adsorption, we consider it different in its application. We therefore think it best to describe it in a different context. You will notice that our applications are environmentally focused.

A natural and seemingly inevitable result of industrial development and human activity is the release of organic and inorganic contaminants to the environment. We, as human beings, consume raw materials and release contaminants, often toxic. Industrial development has led to the release of pollutants that range in toxicity from benign to acute to chronic. Agricultural progress, especially in the control of insects and weeds, has developed its own set of well-known pollutants.

Most of these contaminants are handled naturally by the biosphere. Clays and rocks are able to remove many pollutants from the water by ion exchange and adsorption processes. Bacteria, molds, and algae all have the ability to metabolize most pollutants. Septic tanks and municipal water waste treatment facilities depend on bacteria to degrade human waste. When new pollutants are introduced into the environment, microorganisms, in most cases, evolve to be able to use the contaminant as a food source.

The concentration of population in urban areas and large releases of industrial pollutants in many cases, however, outstripped the ability of the environment to handle the concentrations. Additionally, there is a class of synthetic organic pollutants that has been designated recalcitrant in the sense that the

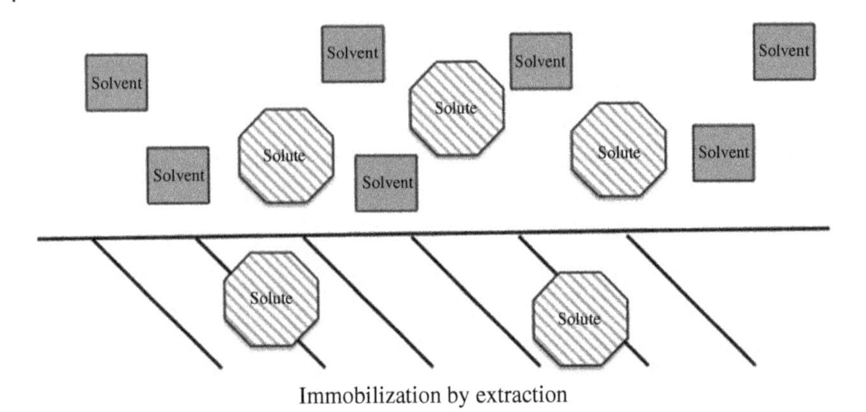

Figure 4.16 Immobilization by extraction.

natural environment does not seem to be able to remove them. Halogenated hydrocarbons and certain pesticides are in this category.

The US Department of Agriculture monitors the pesticide concentrations in groundwater, usually as a result from runoff. The chlorine- and phosphate-based chemicals are among the most toxic releases. An added complication is that these are nonpoint source releases. That is to say, there is not an identifiable release point that can be controlled. In any case, this mode of chemical release contributes to a background of synthetic organic contaminants.

Many of these organic, and in some cases inorganic chemicals, have a potential for bioaccumulation; therefore there is a need for effective means for remediation. Among the processes used to treat or remove them to safe levels are steam distillation, liquid–liquid extraction, adsorption, chemical oxidation, various biological processes, and membrane techniques.

The separation of contaminants by extraction is the focus of this section Figure 4.16. Extraction, however, is most commonly practiced on a liquid-to-liquid basis. That is, two immiscible liquids are contacted, one clean and the other contaminated. If the contaminant is soluble in both liquids, some of it will migrate from the contaminated fluid into the clean liquid, thus separating it. At equilibrium, the relative concentration is called the partition coefficient and its value is a function of the relative solubilities, temperature, polarities, and the nature of the liquids themselves.

Our discussion is concerned with immobilization of contaminants and transferring from one liquid to another cannot be considered immobilization. If the same physical chemistry can be used to transfer a contaminant to a solid, even if an intermediate stage in a process, then it can rightly be termed an immobilization. This is the approach we will discuss.

We begin this with an interesting example. It is intermediate between liquid–liquid and liquid–solid extraction, but more importantly it introduces polyethers as an extraction medium. In a useful review article under the direction of R. D. Rogers of the Department of Chemistry of the University of Alabama, Tuscaloosa [11] is appropriate. Several separation techniques are described that involve the addition of surfactant-like molecules to aqueous solutions containing target molecules. Once the solution was complete, a separation of the phases is accomplished by any of several techniques. For the purpose of this discussion, we will describe the most common, aqueous biphasic separation (ABS). The phenomenon is known to most chemists as salting out. If you add an ionic salt to an aqueous solution of an organic molecule, the solubility of the organic decreases and separates. If that organic was used as the extractant, the target molecule is carried along in the separated phase. The separated organic phase is usually liquid, however, and so it doesn't qualify as an immobilizing system.

While the mechanism is important, it is the medium that is used to affect the extraction that is of particular interest. A complete list is not necessary to make our point. To that end, the following two molecules are of particular interest (Figure 4.17).

You may or may not recognize that these polyols are also the basis of much of the PU industry. The former is the polyol used in hydrophilic PU, a small component in the industry, and the latter is probably the most commonly used in the industry around the world. It is the basis for most polyether PUs. Note, however, that while polyethylene glycol (PEG) is water soluble at high molecular weights, propylene glycol (PPG) is not. In fact, high molecular weight and cross-linked PPGs are hydrophobic.

This is the connection, however, we wanted to make. We know it to be true that if one of these polyols is effective as an ABS, then a PU made from the polyol is also effective. In addition, it is without the complication of "salting out" and separating. To illustrate, we reacted a number of surfactants with a 5% excess diisocyanate to produce a series of PU hydrogels. Pluronics (BASF Corporation, Florham Park, NJ) is a polyol intended for use as PU but presumably will work as an ABS. It is designated as GP-5171 (Carpenter Co, Richmond, VA).

Figure 4.17 Glycols used for biphasic separations.

$HO{-}[CH_2{-}CH_2{-}O]_X{-}H$

Polyethylene glycol

$HO{-}[CH_2{-}CH(CH_3){-}O]_X{-}H$

Polypropylene glycol

Table 4.5 Pluronic and other surfactants.

Polyol	MW	Functionality	Solubility in water at 25°C	% ethylene oxide
Pluronic L62	2500	2	>10	20
Pluronic L101	3800	2	Insoluble	10
Pluronic L121	4400	2	Insoluble	10
GP-5171	5000	3.1	>10	71
Pluronic P123	5750	2	>11	80
Pluronic F88	11400	2	>12	80
Pluronic F127	12600	2	>13	75

As a reminder the functionality controls the three-dimensionality of the resultant PU. And it is the functionality that produced liquid PUs from all of the Pluronics (Table 4.5). Our intent was not to produce anything useful except to demonstrate that PUs could be made from polyols that could be used in the Rogers technique. This was successful. The GP-5171 however made a hard gel when reacted. We discussed these experiments as part of a study to make a spinal disc.

With that rather long introduction, we have made an empirical basis for the use of PUs as an extraction medium. We will present several examples from our research and that of other researchers. In many cases we will compare the results with activated charcoal, the most commonly used extraction system for gases. In those comparisons it is generally accepted that charcoal is more effective but PUs are considered superior in flow-through and regeneration. Also, PUs are more appropriate for aqueous separations and, as we will show, more effective for high molecular weight molecules. It is these properties that differentiate it from activated carbon.

Gesser *et al.* [12] initiated the application of PUF for the extraction of organic contaminants from water. Since then innumerable investigations have been published describing the application of foams as extractors for chlorinated organic compounds, polycyclic aromatic hydrocarbons (PAH), and other organics.

Saxena *et al.* [13] used PU as a system to concentrate PAHs found in water. After passing spiked water through the foam, the amount of PAH was measured. PUF was used as an alternative to activated charcoal for the extraction of the organic contaminants in water. Water passed through the foam column at about 150 ml/min. It was noted that higher flow rates were used compared with the rates that could be used with carbon. According to the author of the study, the PUF was far superior to the active charcoal.

In another study, PU was used to extract phenol [14]. The extraction efficiencies of the foam were compared with other extractants including activated charcoal. The urethane had the highest efficiencies (60–85%). Carbon was 45% effective.

Faudree [15] teaches the production of an alcohol-grafted PU for the express purpose of absorbing waterborne oil spills. By attaching long-chain aliphatic hydrocarbons to a PU prepolymer, they were able to increase the ability of the PU to increase the absorption of 10W30 oil by about 20%. The patent, however, goes further to teach that the PU should be ground into a powder. We improved the technology, we feel, by grafting the Faudree polymer to a reticulated PU to produce a sheet of oil-absorbing foam. In our tests on 10W30 motor oil, we demonstrated its superiority over untreated foam. The visual comparison is seen in Figure 4.18.

While the properties of PUs as an extractant are useful, several problems make it less than ideal. PUs are far more specific than activated charcoal in removing contaminants. As we discussed, charcoal separates by size, and while there is some specificity, this mode of operation is well suited for mixtures of diverse chemistries (PAH vs. halogenated hydrocarbons); PURs however operate on the principle that "like dissolves like." Conventional PUs are made up of a hydrophobic isocyanate and hydrophobic polyalcohols. Thus the molecule is hydrophobic. There is some polarity in the polyalcohol backbone, but it is hindered and therefore has a net low degree of polarity. Inasmuch as the extraction effect is based at least in part on polarity, we would expect (and it has been shown) that PUs are most effective for nonpolar pollutants. An example follows in studies on the use of PU to extract pesticides from water.

Figure 4.18 Improved absorption of oil by a derivatized polyurethane foam (left). *(See insert for color representation of the figure.)*

Extraction of Pesticides

El Shahawi [16] has studied the use of PU for the extraction of pesticides and similar compounds from water. In the 1993 paper, he reported on an investigation of the extraction of Dursban, Karphos, and Dyfonate by activated charcoal and a polyether PUF. A weighed portion of carbon and PU were placed in standard solutions of the pesticides, and after a period of soaking and agitation, the concentration of organic in the supernatant solution was measured. The data is restated from the report as the percent of the pesticide extracted from the solutions. This is seen in Figure 4.19. With all three chemicals the PUF was superior. This is probably related to the molecular weights of the pesticides.

As important, El Shahawi went on to hypothesize about why there is a difference in the degree of extraction. The study employed hydrophobic (nonpolar) polyether PUs for his work.

The reasons included molecular weight but they also concluded that the polarity of the molecule was a contributing factor. They stated, "the smaller the dielectric constant of the absorbate the larger the amount absorbed." Restating this as the polarity of the pesticide increases, the ability of a conventional PU to extract it decreases.

What we hypothesize is that if he had used a hydrophilic PU, he might have concluded that "as the polarity of the pesticide increases, the polarity of the extracting polyurethane must increase." It follows that the broader the polarity range of the target solutes, the more important it is to use a PU with a broad polarity.

To further explore the effect of the polarity of the molecule on its extractability, we studied the extraction of a water-soluble (and therefore polar) dye molecule by polyether PU (hydrophobic) and a hydrophilic foam (HPUR). The dye was bromothymol blue (BTB), a pH-sensitive dye that changes color from yellow to green to blue at pH 6 to 6.5 to 7.0, respectively. Standard solutions of the dye were made and the pH buffered at 8.0. The intensity of the color in the supernatant was monitored by visible spectrophotometry. Samples of the two PUs were added and stirred continuously. If there were no extraction, the % transmission of the liquid would not change. If there were extraction, however, the % transmission would increase as a function of time.

In Figure 4.20 we report the data on a conventional (i.e. hydrophobic) polyether PU. There was little or no extraction of the polar dye molecule.

The effect of a hydrophilic foam, for example, extraction/immobilization, is evident (Figure 4.21).

It is important to note two factors. The hydrophilic foam was permanently stained by the dye, and it still responded to the same color changes in response to pH. Secondly, the molecular weight of BTB is 650, thus supporting the molecular weight hypothesis.

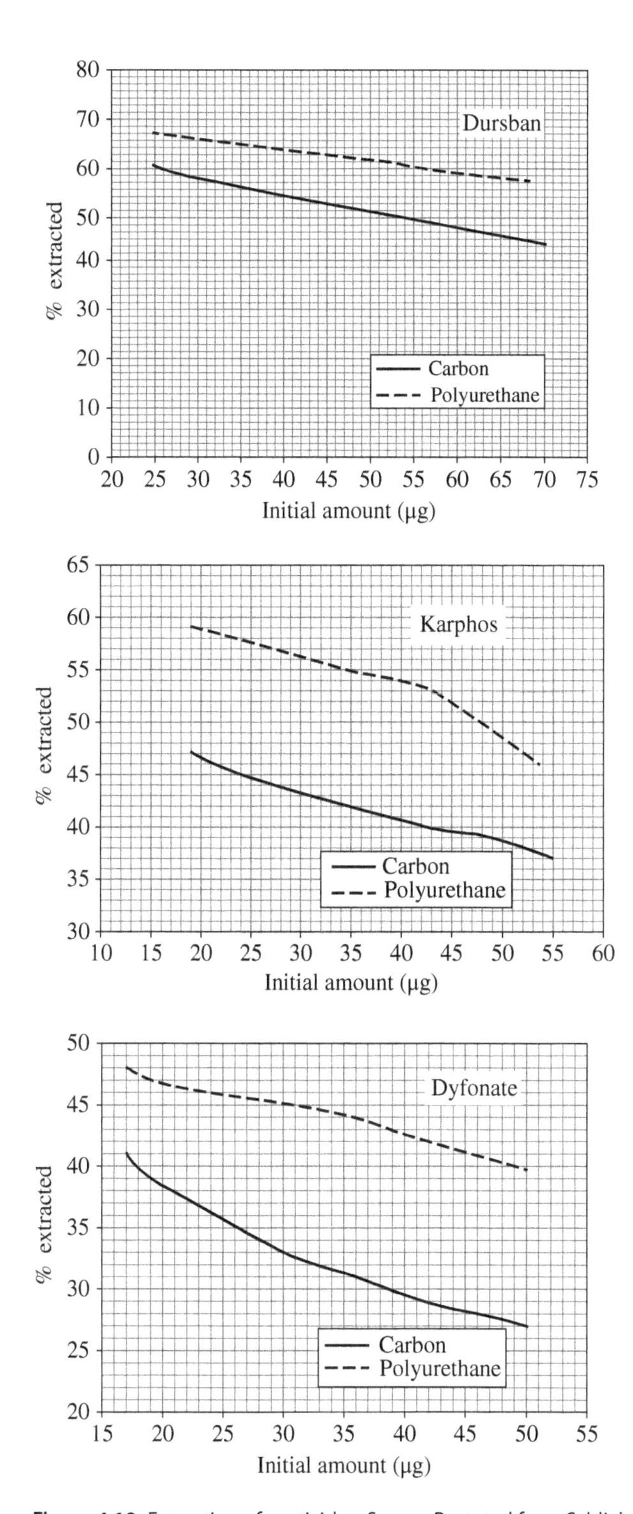

Figure 4.19 Extraction of pesticides. *Source:* Restated from Schlicht and Mc Coy [14].

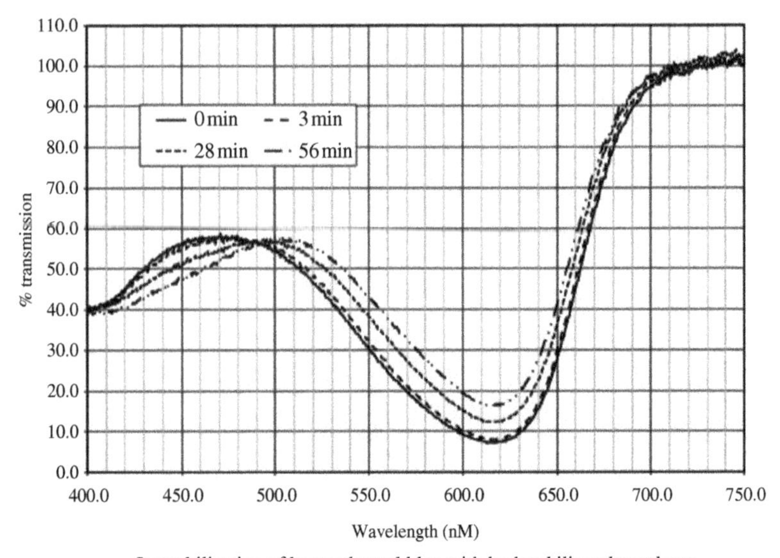

Figure 4.20 Transmission of an organic dye with hydrophobic polyurethane.

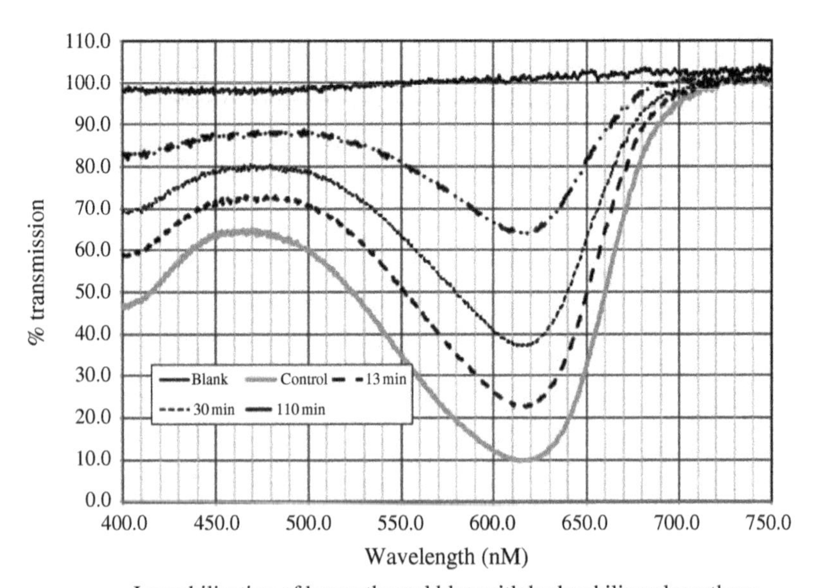

Figure 4.21 Extraction of an organic dye with a hydrophilic polyurethane.

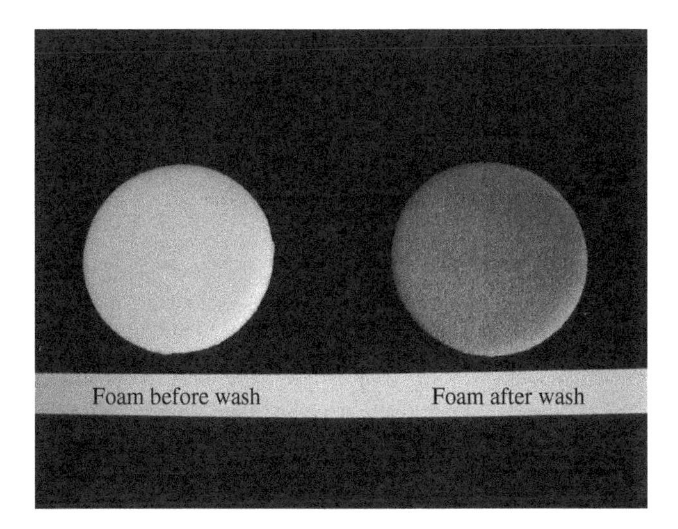

Figure 4.22 Extraction of a dye using hydrophilic polyurethane. (*See insert for color representation of the figure.*)

Our intent here is not to describe commercial ventures but rather to describe the breadth of the immobilization effect. In that context, we were interested in dyes for coloring clothing. The details of the dyes were not known to us, and so we cannot elaborate on the experiment other than the following.

Six extra large men's sweaters (three dark blue, one red, and one green) were washed using the recommended procedures in a household washing machine. We used a commercial detergent. Several pieces of hydrophilic PUF were placed in the washer along with the sweaters.

After a standard cycle the hydrophilic foam was removed and the color inspected. To follow is the results of that study (Figure 4.22).

Based on this dye immobilization phenomenon, we spent a significant time on developing a forensic system for detecting explosive and controlled substances. If fully developed, it could be used for on-site analysis of suspected chemicals. In Table 4.6, we see some of the indicators for, in this case, explosives. A similar table could be assembled for the detection of controlled substances (heroin, for instance). Our purposes are met, however, by using explosives.

Kits are available with these and other color indicators. They typically include bottles of reagents and a plastic plate on which the analysis is performed. Interferences and other ambiguities, however, limit the use to "indications." It was our intention to apply the immobilization technique described earlier to this application. Still further we felt that by including several dyes, isolated to regions on a single foam disc, we could remove some of the ambiguities. For instance, while diphenylamine turns blue in the presence of gunpowder and

Table 4.6 Colorimetric determination of explosives.

Analyte	Diphenlamine	Antazozole	Perchlor A/B	Chlorate	Peroxide reagent
Perchlorates	Colorless	Colorless	Purple/specks	Colorless	Purple
TNT	Blue	Colorless	Blue	Colorless	Purple
Gunpowder	Blue	Pink-red	Blue	Colorless	Purple
Nitrates	Blue	Pink-red	Blue	Faded blue	Purple
Dynamite	Blue	Pink-red	Blue	Red/purple	Purple
Pyrodex	Blue	Pink-red	Purple	Colorless	Purple
Chlorates	Blue	Red	Blue	Blue	Purple

trinitrotoluene (TNT), antazozole is colorless when exposed to the latter. Thus a single foam disc adsorbed with the two indicators would differentiate gunpowder from TNT.

We developed a spectrophotometric technique to analyze treated samples. While the technique was never tested on explosive residues, it was simulated by appropriately colored disc based on known color responses of typical explosive residues. The following apparatus was set up to measure the "average" color of a spinning disc. The spectra were taken using a board-level visible spectrophotometer. By rotating the coupon faster than the response time of the spectrophotometer, an "average" spectrum is collected (we call it a "rotational spectrum") (Figure 4.23).

In the following graphic, synthetic coupons of representative colors are used to test the system. Coupons were made with colors. Figure 4.24 shows the average rotational spectra of coupons of various color segments.

In another example of comparing PU with carbon, a large consumer products company investigated the use of PU to extract "bathroom odors." While carbon is commonly used, apparently they felt there was room for improvement.

They used synthetic substances for that study. Other than the following suggestion, the study was done without our assistance. We had mentioned that we and others had seen a molecular weight effect. Accordingly, they used surrogates of several molecular weights. We provided coupons of Hypol-based hydrophilic PU, and they compared the extraction of the target molecules with activated charcoal (supplied by them). The results are reported here as the efficiency of the PU compared with carbon (100% implies that the efficiencies are the same). The molecular weight effect is confirmed. Above MW = 84, the immobilization is equal or more efficient (Figure 4.25).

Figure 4.23 Rotational spectroscopy using a fiber optics. (*See insert for color representation of the figure.*)

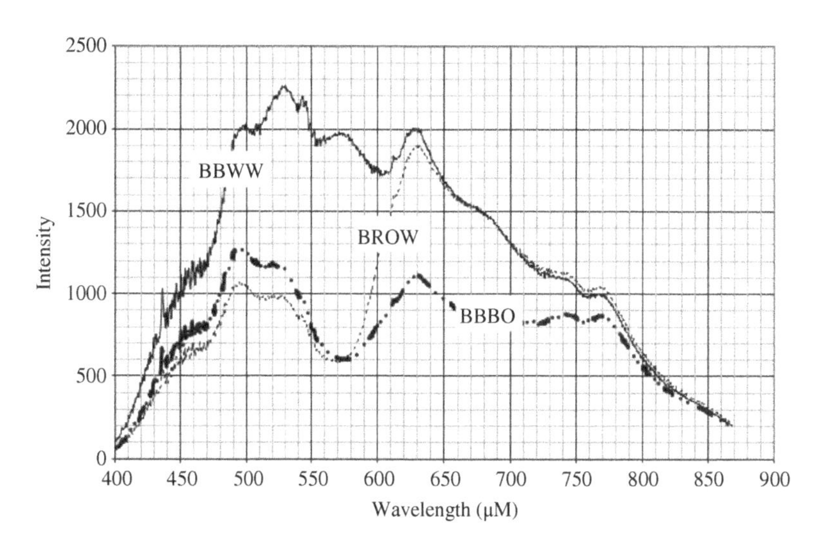

Figure 4.24 Rotational spectroscopy of various color coupons (B, blue; R, red; O, orange; W, white).

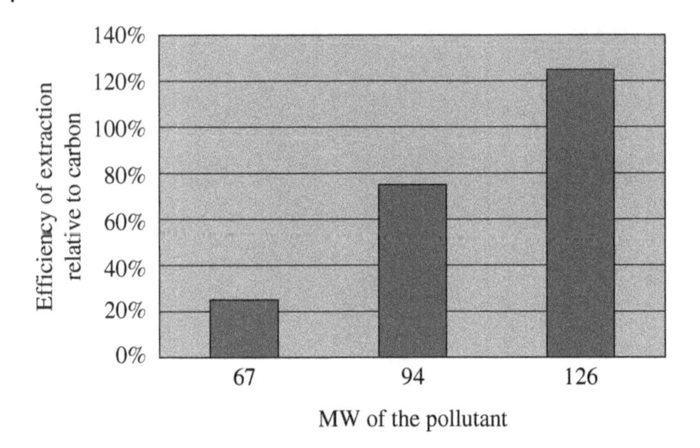

Figure 4.25 Extraction of "bathroom odors" by molecular weight.

The last example, again, shows the breadth of immobilization by extraction. This experiment seeks to differentiate PUR and HPURs; we measured the extraction of polar fragrance molecules from air. As seen in Figure 4.26, we placed coupons of HPUR and PUR foam on the pan of an electronic balance. Inside the chamber we placed a pan containing a fragrance (Avon "Fifth Avenue" essential oil from Shaw Mudge). The mass was monitored for 24 h. In that period, the hydrophobic foam gained less than 1 mg of weight, while the hydrophilic gained 42 mg. This experiment was conducted in support of the use of HPUR as a fragrance delivery system. An important aspect is the length of time that the system would release the fragrance. The odor in the sample persisted for over a month.

Summary

This concludes the discussion of the use of extraction physical chemistry as it applies to immobilization. Our discussion is extraction in nature but has focused on extraction from a fluid (gas or liquid) into a solid, specifically PUs. We have mentioned that the action is a partition effect with the relative solubility being the controlling phenomenon. We compared the immobilization of both polar and nonpolar chemicals using polar and nonpolar PUs. We applied these principles to a broad spectrum of applications. Remember, however, that this is a portion of the immobilization discussion. Still further, this discussion is part of the more general subject of immobilization using a scaffold for support or for mass transport.

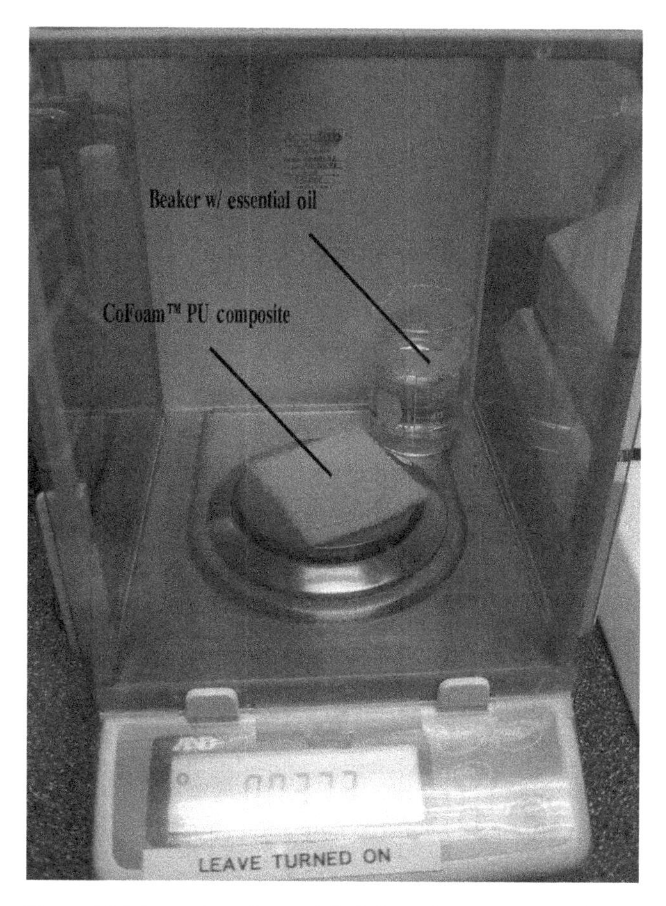

Figure 4.26 Extraction of essential oils from the air. (*See insert for color representation of the figure.*)

Immobilization by Entrapment

The last method of immobilization, unlike the other techniques, is less applicable to a flow-through scaffold. The reason is that it is a scaffold itself. That scaffold as we will show is a permeable membrane, the chemistry and surface area of which control the activity of the biological it encases. Thus it functions as a scaffold in its own right. There is a philosophical reason to include it in our general discussion of scaffold/biomaterial composites. To explain, the natural development of tissue includes, as we have discussed, includes an ECM. While the function of the ECM is complex, it serves as a semipermeable layer around

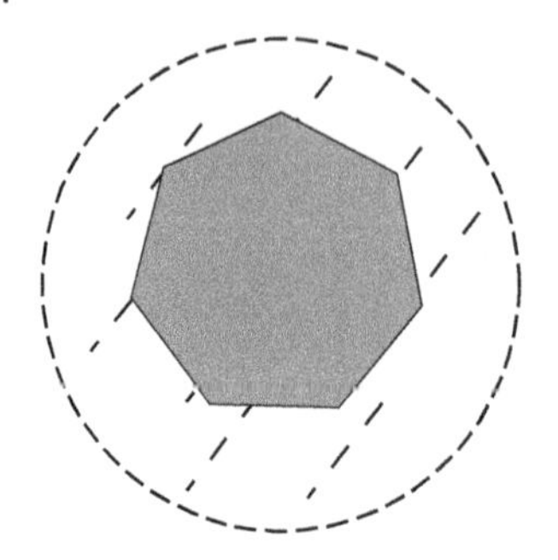

Figure 4.27 Immobilization by entrapment.

the cells. It can therefore be considered an immobilization (at least for this discussion). We have discussed the adsorption and covalent attachment of cells and biomolecules to a flow-through scaffold. While we will discuss the encapsulation of biologicals without regard to flow-through characteristics, one should keep in mind that the techniques taught in this section apply to the encapsulation of an entire colony of cells or biomolecules within the structure of, for instance, a reticulated PUF. A construction of this type would mimic, at least in part, the natural tissue construction to coat and protect an entire colonized scaffold with an encapsulating membrane (cells encapsulated in an ECM) (Figure 4.27).

This technique like the others discussed so far effect the activity of the bioactive, but it can be useful.

Immobilization by encapsulation like the other immobilization techniques improves the durability of the biomolecule or cell. While the process negatively effects the activity, in general, the compromise favors immobilization. One of the common methods to affect the benefits of immobilization is by encasing the cell or biomolecule (an enzyme, for instance) in a semipermeable membrane. Typically, but not always, the membrane is a cross-linked hydrogel. Clays and silica-based materials are also used. In the hydrogel category are acrylamides, polyvinyl alcohol (PVA), and a number of naturally occurring materials (pectin, alginates, gelatin, etc.).

Unlike many of the covalent immobilizations we will discuss, activation of the surface is not necessary as the first example illustrates.

Alginate Encapsulation

Entrapment of an enzyme or cell in an alginate is a convenient place to begin with this review. Alginates are commercially available and are used in food and pharmaceutical formulations [17]. In the following example yeast cells are incorporated into an alginate gel. The immobilized yeast could be used to convert glucose into ethanol. A simple experiment illustrates the technique.

An appropriate amount of sodium alginate is dissolved in water to yield a 3% alginate solution. The wetted yeast is added with careful mixing to avoid bubbles. The yeast–alginate mixture is added dropwise into a solution of 0.05 M of $CaCl_2$ with continuous mixing. Gel formation develops as soon as the sodium alginate drops come in direct contact with the calcium solution. The size of the beads is controlled to some degree by the aggressiveness of the mixing. The beads will harden in 1–2 h at which time they can be used to metabolize the glucose.

We have focused on the protective nature of encapsulation. The function is characterized in several ways, but the physical coating of fragile proteins would appear to be the primary mechanism. We now know, however, that the mechanism is more complex. To explain, we have discussed the morphology of enzymes and its importance in orienting the proteins. The active segment of the enzyme must approach the target in a specific way and be stabilized by adsorptive processes. A recent paper suggests a synergistic effect that contributes to the stabilization [18]. The morphology of enzymes is maintained by the water in which it exists. If the water is removed, the structure of the molecule is destroyed. If the water exceeds 40°, the activity of the water changes enough to separate from the protein in the process called denaturation.

In their study, the authors explained that traditional experimental conditions don't mimic the thermodynamics and kinetics of biological processes in a living cell. For instance, the rates of protein folding and unfolding are not the same when done *in vitro*. The reason was, in part, due to be based on the association of the hydrated encapsulant and the protein. In the study they referred to the stabilizing effect as "molecular confinement."

The agarose gel is a method of immobilizing water. When the hydrated polymer comes into contact with a protein during encapsulation, the immobilized water has the effect of stabilizing the hydrated protein structure. The agarose backbone serves as rebars (our characterization, not the authors). Their study indicated that confinement in agarose gels limits the flexibility of the polypeptide backbone, thereby preventing protein unfolding at higher temperatures, for instance.

In their study, they investigated the effects of molecular confinement on enzyme kinetics, as well as their functional and structural stability under denaturing environments. Agarose was used to encapsulate the enzymes of interest. They used two different concentrations, 0.5 and 2%, to encapsulate the enzymes. The enzymes were encapsulated using standard entrapment techniques [19]. The model enzyme was horseradish peroxidase (HRP).

Enzymes encapsulated in 2% gels were significantly more stable relative to enzymes encapsulated in 0.5% gels and solution phase enzymes under denaturing environments (ethanol exposure). This supports our hypothesis that the agarose has a reinforcing effect on the conformation of the enzyme.

Encapsulation of Pancreatic Islet Cells

Insulin therapy is the mainstay of treatment of type 1 diabetes. However, this therapy provides only temporary resolution and cannot fully prevent the morbidity associated with chronic complications. Transplantation of pancreatic islets has received research interest as a potentially curative treatment [20]. Realizing this potential depends on effective methods for isolating and preserving islets, assuring immune protection, and providing stable placement of islets at the proper position in the body.

The first of these obstacles is the isolation of islets while preserving a well-oxygenated environment and maintaining the supply of nutrients. In an *in vivo* environment, islets form capillary ball-shaped clusters. During islet isolation, however, these microvascular structures are destroyed. The isolated islets are not properly oxygenated or supplied with nutrients. Approximately 50% of transplanted islets are lost during transplantation period, mainly due to the hypoxic state of large vessel-free islets at the transplanted site [21].

One method that may be suitable for overcoming this obstacle is through encapsulation within hydrogels [22]. In a recent study [23], an alginate and collagen–alginate encapsulate was evaluated for islet viability, oxygen consumption rate, and insulin secretion. The encapsulation was performed in a concave microwell array containing the spheroids. The encapsulated islets were implanted into the intraperitoneal cavity of mice and evaluated for glucose control. Intact islets were also transplanted as control to investigate the effect of encapsulation.

The researchers concluded that tests of glucose secretion and oxygenation ability showed that the collagen–alginate encapsulate was an excellent encapsulating material that permits good gas and nutrient exchange. The encapsulates maintain glucose levels of less than 200 mg/dl and showed a strong immune-protective function.

This lays the groundwork for a device that includes the development of spheroids within a reticulated PUF by methods developed for hepatic cells [24]. Certain adsorption methods discussed in this chapter could be used to secure the cells to the scaffold. This would be followed by an encapsulation of the internal surfaces of the cell colonized scaffold.

Encapsulation of Osteoblasts

Hydrogels are often considered as encapsulants. The high water content (as much as 95%) makes them inherently biocompatible. This is especially true when the hydrogel is swollen with biological fluids. They exhibit high permeability for oxygen, nutrients, and other water-soluble metabolites [25]. For instance, the Pluronic family of surfactants has found acceptance as an encapsulant due to its amphiphilic nature. It will form a physically cross-linked gel. Pluronic® F127 will undergo a gel transition at a concentration higher than 14 wt% and a temperature of 30°C [26]. The gels, however, are relatively weak and must be strengthened if they are to be considered for implantation. A method for making the gels more durable was developed by converting the hydroxyl end groups to *N*-methacryloyl-depsipeptides [27]. You will remember that we converted many Pluronic surfactants into solid gels by using a cross-linking polyol and an isocyanate.

In a study that employed this technique, the derivatized gel was used to immobilize osteoblasts. However, osteoblasts are anchorage-dependent cells that need a scaffold. For this reason, cells were first cultured on gelatin beads and then mixed with a modified Pluronic.

Cell survival and osteogenic differentiation were evaluated in a standardized goat model. The gel–microcarrier mixture was applied in a defect in the long bone of the leg of the animal. About 2 weeks after implantation, the cell viability of the osteoblasts was microscopically assessed. The study indicated homogeneous spreading of the cells, resulting in regeneration.

Again we see the synergy of cell and scaffold. In this example, the scaffold was made of gelatin beads. One would expect that a gelatin-coated reticulated foam would also be effected.

There is a special case of encapsulation with which we want to conclude this section. It is special because of its importance and that it introduces an important aspect we have not discussed. In this sections as we did with the liver model in the last chapter, we want to focus on a specific example, not as a goal, but as an example of how this technology might be used. Thus in this section we introduce a pancreas model.

Introduction to the Pancreas Model

In our discussion of scaffolds, we introduced the concept of a "liver model." It was, rather than a suggestion, a tool that allowed us to focus on the function and components on a scaffold appropriate to that need, that is, for an extracorporeal liver assist module. While physical in nature, biocompatibility was discussed. Implied, if not recommended outright, were the possible benefits of reticulated PUF. To this, grafting any number of polymer systems to give it the preferred hydrophilic surface was suggested.

In this chapter, we have focused on how one might augment the function of a scaffold by immobilizing cells of biomolecules on the surface of a scaffold. The various techniques of immobilization, whether as a surface treatment or imbedded within the scaffold or composite, in a sense, activated the construction. Not included specifically in those thoughts was the concept of controlled release of useful chemicals. We will finish this book with an introductory chapter on controlled release (not a part of our wheelhouse) focusing on the control of the diffusion coefficient of polymer including PUs. While we will not discuss PUs directly, as we did in other sections, the PEGs have been investigated in this regard and we think that research is appropriate. Our purpose here is to take a wide view of the subject and by implication show how PUs might play an important role. Our focus will be on research into the search for a bioartificial pancreas. We will use this research again when we talk about controlled release and diffusion, but for now let me introduce the subject with the immobilization by entrapment of the islets of Langerhans. They are the regions of the pancreas that contain its endocrine-producing cells, including insulin. In the scaffold chapter we introduced that "liver model" not so much as a suggestion but rather as a vehicle to focus our intentions on the critical factors that are important to achieve the goal of an extracorporeal device, a liver support device if you will. We will use the same vehicle here for the same reasons, but as you will see, that task has focused on encapsulation of islet cells. It is therefore better to include in as an immobilization system. That is not to say that an appropriate scaffold is critical. The architecture of the scaffold for a bioartificial pancreas is most probably similar to that for the liver, but the technical challenge seems to be in the encapsulants. It must have several characteristics, including protecting the islets from assault. A critical factor is diffusion in and out of the capsules for the islet. In the final chapter we will give examples on how diffusion is controlled. For now however we will review the advantages and problems with the immobilization of islet cell by encapsulation.

The Pancreas Model

Diabetes affects more than 150 million individuals worldwide. In the United States 29.1 million people or 9.3% of the population have diabetes(American Diabetes Foundation web site). Complications include cardiovascular diseases, renal failure, amputations, and blindness. The increasing prevalence and cost of diabetes is driving innovative research projects at the frontiers of medicine and bioengineering,

Therapies for diabetic patients at early stages include insulin injections and dietary restrictions, but therapies for diabetic patients with severe symptoms involve transplantation of the entire pancreas. Short of this drastic measure with its associated complications is the concept of injecting host or xenographic islets into the pancreas. Islet transplantation, as it is called, involves the transfer of healthy islet cells from a donor to the patient. Although generally successful, it is not without complications. Among them are the availability of host islets, preserving islet functions *in vitro*, and rejection of implanted islets. As with whole organ transplantation, patients are required to take immunosuppressive drugs. Still further the source of islet is problematic, but also the expansion of a colony of cells presents problems not unlike the problem of culturing hepatic cells *in vitro*. Quoting from Beck [22],

> The major obstacles to islet transplantation are the availability of islets and the maintenance of islet functions such as cell growth and survival. For instance, islet cells, unlike other cell types, cannot be expanded in vitro to provide sufficient cells for transplantation. Islet cells also tend to clump together, causing the core cells to die because of the limitation of nutrient transport to the aggregate center, which subsequently reduces cell functional replacement.

We now know that the agglomeration of cells is similar to that of hepatic cells. That is to say that spheroids are the natural conformation of islet cells and to limit their natural development on flat plates does not allow them to function properly, as is the accepted philosophy in hepatic science.

We will return to the culturing of islet cells shortly, but let us turn out attention to the implanting of the cells with its problems. Rejection is, of course, the immediate concern. Discussed in some detail earlier in this chapter is the widely practiced immobilization technique of encasing the cells in a hydrogel-like matrix. Islet encapsulation uses a biocompatible biomaterial to create a membrane around a spheroid of islet cells. The membrane allows the diffusion of insulin without inhibiting nutrients and oxygen reaching the cells. Isolation of the islet cells from the immune system may make xenotransplants possible.

As we said, a primary function of the encapsulating material is to protect the cells from assault by the immune system. As a practical matter, these encapsulates are hydrogels with high water content. Not any hydrogel will due however as diffusion in and out is critical. We discuss diffusion and how one controls it in the next chapter, but for now let us review some of the materials that are under consideration.

Alginates are the most widely used for encapsulation of islets. They are, in fact, in clinical trials [28]. While alginates seem to satisfy the current investigative needs, improvements are under constant investigation. We have discussed encapsulation of cells earlier in the chapter, and so it is not useful to repeat it here with the exception of projects that would appear to link the subject to PU and PU derivative molecule. The first example illustrates.

Cui *et al.* [29] demonstrated that grafting PEG chains onto alginate capsules increased *in vivo* viability of islet cells [30]. Our purpose in this discussion is to describe the current status of encapsulation and perhaps add to the technology. We will, of course, discuss the possible role of an appropriate scaffold as a way to increase surface area. In this regard we will discuss our work in flow-through systems as described in our liver model. You have already noted that our now familiar PEG is used. This is our gateway into introducing PU into the catalog. In Cui, many researchers used it in its photosensitive form (adding an acrylate end groups) while we use diisocyanates. Islets were isolated and subsequently encapsulated in 2% alginate. After coating with a polyelectrolyte multilayer of polylysine and alginate, a polymeric coating was applied to the capsule surface.

In a paper by Weber [31], the diffusion of insulin from islet cell encapsulated in a series of PEG hydrogels is measured. Islets from adult mice were cultured at 37°C in humid conditions with 5% CO_2. Approximately 20 islets were suspended in 30 μl of hydrogel precursor solution followed by photopolymerization. Cell viability was determined by staining. Insulin secretion was evaluated by exposure of the encapsulated islets to glucose. The insulin content of each solution was measured, and the insulin secretion was plotted as a percentage of the total amount secreted (Figure 4.28).

As with hepatic cells, islets seem to require 3D culturing. Their natural conformation, as with liver cells, is into spheroids [32].

As we saw in Beck, the conventional culturing of islets results in agglomeration and apoptosis. We hypothesized a connection to what occurs with hepatic cells unless 3D culturing methods are used. To illustrate we can begin to discuss the useful technique of co-culturing islets with hepatic cells. It was discovered that both cells are benefited. We will discuss this in paper on culturing of islet cells [33].

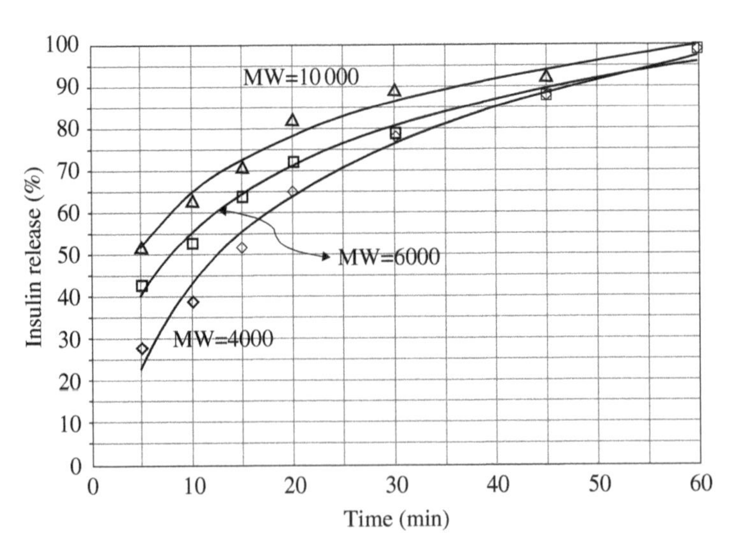

Figure 4.28 Insulin release from islets encapsulated in PEG gels. *Source:* Restated from Weber et al. [31].

Islets and hepatic cells were isolated from rats. The cells were then dispersed to single cells. Spheroids of each were generated in *concave molds.* The concave micro-molds had 400 mm diameter microwells. Cell aggregation and spheroid formation were observed microscopically.

The functional assessment for each cell line was determined individually. A glucose-stimulated insulin secretion assay was used to assess the response of spheroids to varying concentrations of glucose. The amounts of insulin secreted into the low and high glucose solutions were measured. Secretions of albumin and urea from the hepatic cells were also determined. With regard to the latter, you are reminded that it is generally accepted that in order for hepatic cells to be viable in their metabolic function, they must be cultured in such a way as to allow them to develop spheroids. Cells grown in 2D monolayers do not develop the normal cell-to-cell communication that define their identity and function [34]. *In vivo* cells are supported by a 3D architecture, which allows for cell–cell communication by direct contact (Figure 4.29).

Compare this with spheroids developed within a reticulated PUF (Figure 4.30).

Kang *et al.* report that their 3D co-culture system for islet cells and hepatocytes using was appropriate. They concluded that the formation of co-cultured spheroids resulted in high stability and higher viability compared with cultures of the individual cell lines. The co-culture 3D models showed hepatocyte-specific functions, including albumin secretion and urea secretion, even though 75% of the cells were islet cells. This proposed model could potentially permit cell-based therapies for both diabetes and liver disease and "could be further developed into a bioartificial pancreas or liver."

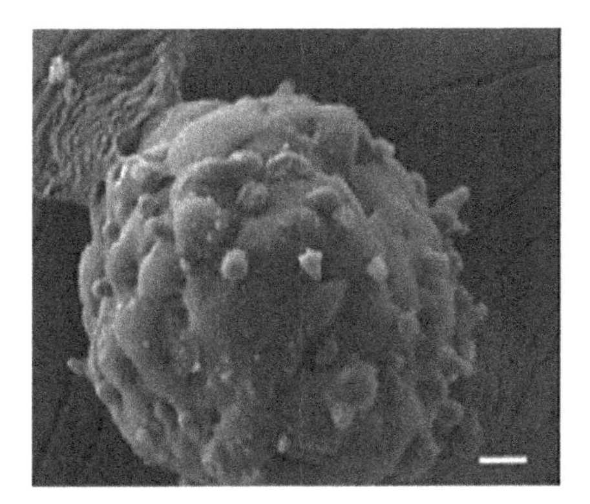

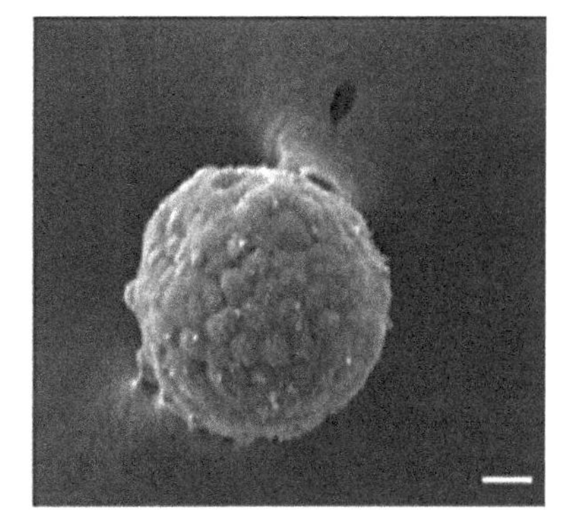

Figure 4.29 Comparison of hepatic (left) and islet spheroids cultured in a 3D environment.

Figure 4.30 Hepatic spheroid cultured in a reticulated polyurethane foam (100 μm). *Source:* Matsushita *et al.* [24]. Reproduced with permission of Elsevier.

Summary of Encapsulation

While the examples discussed in this section do not deal with the kinds of scaffold that we discuss, the use of scaffolding in general is confirmed. Thus it is a big jump to imagine applying the techniques taught in this section to a flow-through scaffold. That is to say, there could be an advantage in some cases, both medical and environmental, if an encapsulate were grafted to a flow-through structure. We know, for instance, that bacteria for a polysaccharide gel (called slime) are presumably for protection.

Immobilization by Covalent Bonding

A preferred method, while not always possible, is to create a covalent bond between the bioactive group (enzyme, cell, etc.) and the scaffold. By this technique, an exchange of the electrons takes place, producing a chemical change in both the scaffold and biological to be immobilized. We will discuss some of the methods researchers have used to affect such a bond. Whether for medical devices or an environmental system, the covalent bond is the most secure attachment. We will use examples of covalent bonding of both cells and biomolecules. We will also show how one can attach other active molecules including ion exchange moieties. In a special section we will examine the attachment of ligands that increase the adhesion of cells to scaffolds (Figure 4.31).

Once the material for the scaffold has been selected, a method of covalent attachment must be developed. It is important to note that some of the methods to attach an enzyme, for instance, are quite harsh. The durability of the scaffold must be taken into account. Nevertheless, designing the scaffold first and then developing an attachment strategy is typically more efficient. This is

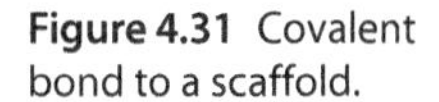
Figure 4.31 Covalent bond to a scaffold.

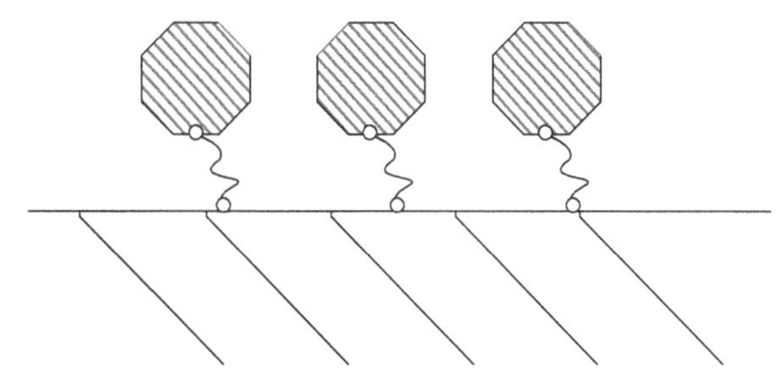

especially true with PUs. At the end of this section, we examine the special case of covalent bonding with PU. We will show that it is the simplest and most efficient method of covalent bonding to a scaffold.

Covalent bonding is differentiated from adsorption by the method of attachment. While adsorption, as we have discussed, is an electrochemical effect involving polarities and other factors, covalent bonding involves the co-mingling of electrons. In addition, adsorption is more or less spontaneous, while covalent attachment involves activation of the surface and a subsequent chemical reaction. In other words, covalent bonding involves synthetic chemistry and typically specialized equipment. We will illustrate this when we discuss the immobilization of an enzyme on cellulose. In any case, all this will be clear when we present some illustrative examples.

Before we begin, we need to make another point. The overall intent of this book, other than to promote PU, is to show an intimate connection among a scaffold and a surface treatment. As we have often said, the future of tissue engineering is the simultaneous development of these two radically different technologies. In the final chapter, we will add a third technology, "controlled release." As we proceed, it will seem off target to discuss much of the research on immobilization without mentioning scaffolding. In many cases the immobilizations are done on flat plates or encapsulated into or on beads. We would advise the reader to imagine the process as it would be applied to a 3D scaffold. The following example illustrates the point we want to make. In principle it applies to any of the immobilization techniques regardless of the system. As we discussed, a common method used to immobilize a biological is to encapsulate it into alginate beads. The process begins by dissolving sodium alginate in water. The biological to be encapsulated is mixed with the alginate while adding a solution of a calcium salt. Calcium alginate is insoluble in water, and so, depending on the rate of mixing, beads of encapsulated biological result. In the broadest sense this is a valid immobilization to a scaffold. In the last chapter we compared a bead-type scaffold with a reticulated scaffold and showed the advantages of what we called the "antibead" architecture. For now however, imagine the alginate solution containing the biological being applied to the internal structure of a reticulated foam and the insolubilized by the calcium salt. By this technique, one would accomplish the goal of a flow-through medium containing an immobilized enzyme. In this simple example we want to imbed in the mind of the reader that the generally accepted need for three-dimensionality with interconnected cells and effective mass transport properties is available to most if not all immobilization techniques. Still further, we tried to make the case that reticulated PUF might be the "ideal scaffold."

Overview of Covalent Immobilization

The subject of immobilization to a surface has more to do with the substrate than with the biological. Presumably, the researcher or engineer knows what is to be immobilized. Perhaps it is an enzyme with known activity. The goal of the project however is to improve the useful life of the biomolecule even at the expense of activity, without regard to the scaffold. What we will be discussing therefore is to change the surface that is to be used without regard to architecture. Therefore the focus is how a broad spectrum of surfaces can be activated so that an effective immobilization is accomplished.

We will begin our discussion with the most common research aspect, the immobilization of enzymes. The technologies apply to other biologicals such as antibodies and cell adhesion ligands, but by studying the enzyme application, we sufficiently describe the immobilization of any biomolecule or cell. The driving force for enzymes is economic. For other biologicals it may be function. In any case, as we have said, this section deals with how the surface is prepared. Once accomplished and combined with an appropriate scaffold, the composite can influence metabolic processes and other aspects, including adhesion, spreading, proliferation, and differentiation.

We will discuss a number of surfaces and the techniques to immobilize active molecules, all in preparation for a summary of the advantages of PUs. The benefits are physical in nature as we have discussed, but in the context of immobilization, we will introduce numerous advantages, not the least of which is that you don't need a chemist or specialized equipment to do the research. What you will learn in the next discussion is sufficient to begin your study of immobilization using PU chemistries.

It is important to review the research with regard to materials that are considered useful in the development of an immobilized system. Many of these are considered biodegradable (an unfortunate philosophy). The reason is that the vast majority of immobilization research is on substrates other than PUs. The discussion will focus on the technique of immobilization and has little to do with the substrate. If immobilization is considered effective, then it follows that immobilization on another substrate would also be of benefit. Thus it is appropriate that we examine a broad spectrum of immobilization chemistries.

Polymers have been widely used in these biomedical applications. Selecting the materials is typically based on chemistry, hydrophilicity, physical strength, the ability to immobilize biomolecules, and *in vivo* degradation. The last characteristic includes the rate and the products of degradation. These and other properties have made polymers a candidate in tissue engineering and organ substitution. An overview of various biomedical applications of polymer-composite materials is reported in Ramakrishna *et al.* [35].

Biomaterials, a general term for the systems we will be discussing, are defined as a material intended to interface with biological systems to evaluate, treat, augment, or replace any tissue, organ, or function of the body. Philosophically, the substrates to which we will immobilize a biological may or may not be a biomaterial. PU is a good example. Once a biological is attached, it becomes, by definition, a biomaterial. The definition has evolved as more and more research has focused on tissue development.

There are three general categories of biomaterials: ceramics, synthetic polymers, and natural polymers. All of these have been investigated as potential substrate materials. As you might expect, each has advantages and disadvantages, so research on the use of scaffolds is becoming common. We will discuss examples of each.

We will examine the materials from an outcome perspective. Our immediate goal is the preparation of a biomaterial with a definable biological significance regardless of whether it achieves an *in vivo* purpose. The reduction to clinical or environmental practice is achieved once the immobilization is applied to an appropriate scaffold.

Substrates Used for Immobilization

Cellulose is the most abundant natural polymer and has been widely used as a substrate for immobilization. It is a polysaccharide consisting of a linear chain of glucose units, each of which has three hydroxyl groups. One of the hydroxyl groups is hindered so it is not available for reaction. The other two, however, can be derivatized. Methylcellulose is produced by swelling the cellulose in a sodium hydroxide solution and then reacting it with methyl chloride. The result is a water-soluble polymer, methylcellulose. To use cellulose as an immobilization substrate, another method is used. One common method of activation of the polysaccharide is by treatments with cyanobromide (CB) and triethylamine (TEA). To use cellulose as an immobilization, substrate activation is required. A procedure is suggested in Paterson and Kennedy [36]. The CB is dissolved in an acetone/water solution cooled to −15°C. The solution is added to the washed cellulose and the system is maintained at temperature (Figure 4.32).

The activated surface can then be used to immobilize proteins. In the reference, an amylase enzyme is dissolved in a carbonate buffer at a pH of 9.5. The solution is added to the activated cellulose. After 2 h the material is washed of excess enzyme and stored in a phosphate buffer at 4°C.

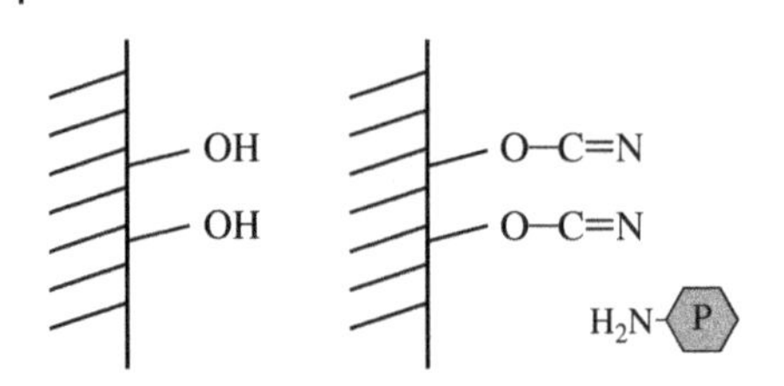

Figure 4.32 Activation of cellulose.

One of the points we will continue to make is the advantage of PU chemistry in this regard. This is self-serving. The aforementioned chemistry leaves you with a cellulosic in sheet or powder form. The process to get there was by the use of highly corrosive and toxic materials that are best handled by trained chemists, not those skilled in engineering or medical device or tissue development. As we will see in many of the immobilization techniques, scaling the aforementioned process into engineering size is beyond comprehension. PU technologies are not limited in this respect. That is not to say, the discussion is not useful. If it encourages research in the area, then it increases the interest in enzymatic processes and then it is a valuable activity.

We asked that you keep in mind that regardless of the substrate, it can be part of a composite. If the researcher prefers a particular substrate from a chemical point of view, but the system is inappropriate in an application in which a fluid can flow through it, one should consider "immobilizing" the immobilized protein onto a more appropriate architecture. Cellulose is an interesting example of how this may be done.

One of the oldest synthetic fibers is based on cellulose. Viscose or rayon is made by dissolving cellulose in a strong base. The result is a highly viscous solution that could be used to coat a surface. That surface could be a polyurethane foam. The cellulose can then be derivatized as described earlier.

By way of example of an immobilized system, consider the following example using another natural polymer, calcium alginate.

Alginates

As we discussed earlier, cells can be immobilized in a calcium alginate gel. The cells are dispersed in a sodium alginate solution. A calcium salt is added with stirring. The calcium salt of alginate is insoluble in water, and so the cell becomes encased in the resultant gel. Also as we have mentioned, while the cells are viable, mass transport through the gel particle architecture limits productivity.

Swalsgood and Passes [37] reported on their research to immobilize the gel onto an alternative substrate, a steel mesh. The process involved "dipping" a steel mesh into a 2% calcium chloride solution and then in a sodium alginate–cell suspension. After a couple of minutes, the mesh is returned to the $CaCl_2$ solution to complete the immobilization of the cells immobilized on the mesh.

Albumin

We have tried to use examples of materials that are used as substrates for immobilization themselves but that through certain techniques can be immobilized on a scaffold that would improve the outcome. In this example we step a little closer to this philosophy by covalently binding a PEG to albumin. This technique is reported to increase stability and function of other proteins [38]. Linking albumin to PEG involved activating the glycol with 4-nitrophenyl chloroformate. The attachment of the activated glycol to the albumin is by a urea linkage, the same linkage that develops using an isocyanate-capped PEG. The later is commercially available from a number of manufacturers.

Collagen

The cyanobromide (CB) activation method described earlier was used to covalently link collagen molecules onto the surface of cellulose and PVA films [39]. The amount of bound protein was greater for cellulose. In other experiments, *p*-toluenesulfonyl chloride was used to activate the cellulose film followed by grafting the collagen as well. The cellulose film, however, became brittle and weak after activation with CNBr, while the PVA films remained flexible.

Synthetic Polymers as Supports

We will now shift our attention away from biologicals to synthetic polymer systems. Of course, PUs fall in this category, but we will discuss them later. Many of the polymers we will use as examples for this section were covered in the chapter of scaffolds. As we did for natural polymers, we will choose only a few to illustrate the activation process. We will begin by describing the activation of two of the most common polymer systems, PVC and polyethylene (PE). These are both hydrophobic, and so it is appropriate to supplement the list with two hydrophilic polymers, (poly)lactic acid (PLA) and (poly)glycolic acid.

Polyethylene

Immobilization of biomolecules onto inert polypropylene (PP) or PE surface is problematic because of the absence of any chemically active functional groups on the surface. Nevertheless, its availability and relatively low cost make them an attractive substrate for the immobilization of biomolecules. It is necessary, however, to "activate" the surface—that is to say, incorporate binding site on these otherwise inert polymers.

Several methods to accomplish this are discussed in the literature. We will mention a few and then go into some detail on one in particular. All of these methods are relatively tedious and time consuming and typically require a person with above average chemical skills. When we discuss immobilizations on PU, we will make the case that most biomedical researchers or engineers have sufficient skills without further training.

Radiation and Plasma Surface Treatments

Typical methods for activating of PP and PE surfaces use some form of radiation or plasma technique. For instance, activation is developed through exposure to gamma radiation. Lofty *et al.* reported on the gamma irradiation of PE beads [40]. The beads were dispersed in a monomer solution of acrylamide and acrylic acid. The mixture was then exposed to 6 kGy from a cobalt source. Proteins were subsequently immobilized on the activated surface.

PP and PE surfaces can also be activated by introducing hydroxyl or amino groups by a plasma technique. Studies have been reported using various approaches and process gases. The relevant studies focus on the applicability of plasma as interfacial bonding layers for the subsequent immobilization of biomolecules. In a 2006 review article, surfaces with carboxy, hydroxy, amine, and aldehyde groups [41] were reported. These groups are considered the main chemically reactive groups useful for the covalent immobilization of biologically active molecules.

Plasma-produced reactive surfaces with amine, carboxy, hydroxy, and aldehyde groups have been used by many scientists because of their compatibility with well-established chemical reactions for grafting of bioactive moieties such as enzymes, antibodies, proteins, and glycosaminoglycans. Such interfacial immobilization should satisfy a number of criteria:

- The linkage should be a covalent bond and therefore stable enough for additional processing.
- The binding site of the bioactive molecules must not interfere with the biologically active area.
- The active sites of bioactive molecules must be available to the target.

Amine-containing surfaces have been prepared mainly by ammonia plasma treatment. They can be developed over a relatively broad range of plasma parameters. Amine plasma polymers are useful surfaces as such for cell colonization, but the large majority of reports utilize them as chemically reactive platforms for the covalent immobilization of biologically active molecules. After treatment, the reaction between the amine-activated surface and carboxy groups on the molecules leads to the formation of amide bonds.

Examples of bioactive molecules that have been successfully immobilized onto plasma-prepared amine groups are DNA, proteins, hyaluronic acid, heparin, immunoglobulin G, enzymes such as glucose oxidase and glucose isomerase, lysozyme, and polysaccharides such as dextran and carboxymethyl dextrans.

Poly L-Lactic Acid

PLA is a biodegradable thermoplastic polymer derived from, among other sources, cornstarch. In tissue engineering applications, it is used because of its degradability. It is often compared with polyglycolic acid. They have similar degradation rates *in vivo*.

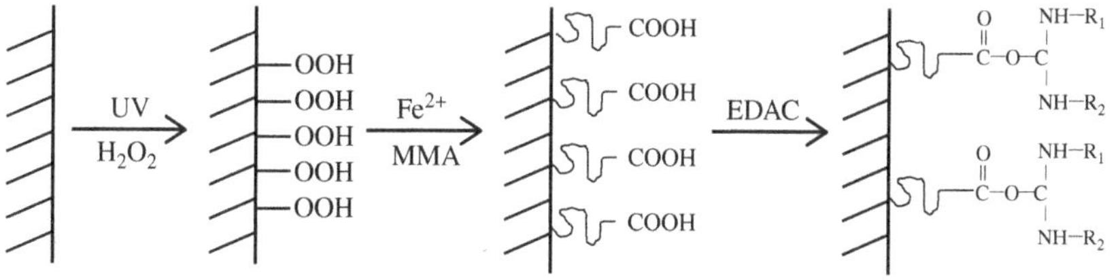

Figure 4.33 MMA grafting and activation of PLLA.

In a remarkable paper, researchers at Zhejiang University, Hangzhou, China, described immobilization of collagen on a 3D PLLA scaffold.

The scaffolds were prepared using a paraffin sphere leaching method [42]. In a practice that we would urge all researchers to use, they described the architecture. The scaffold had a pore diameter of 280–450 μm and porosity of 98%. While porosity is useful, a more appropriate number would be mass transport. Porosity is a measure of void volume but does not indicate the interconnectedness. To continue, however, the scaffold was immersed in 30% H_2O_2 and irradiated under UV light at 50°C for 40 min. The treatment caused –OOH groups to form on the surface as seen in Figure 4.33. After the photooxidation reaction, the scaffold was rinsed to remove the excess H_2O_2. The scaffold was then immersed in methyl methacrylate (MAA) solution and purged with nitrogen. A ferric iron salt was added under constant stirring and the grafting proceeded. The scaffold was rinsed with water to remove unreacted monomer.

The scaffold was considered to be PMAA-grafted PLLA scaffold and this was confirmed by spectrophotometric analysis (Figure 4.33).

The collagen immobilization was performed by first immersing the scaffold in 1-ethyl-3-(3-dimethylaminopropyl) (EDAC) solution for 4 h at 4°C. It was then reacted with a collagen solution (0.3 vol% in acetic acid) for 24 h at 4°C. The scaffold was then dried under vacuum.

To introduce the growth factor (GF), an acetic acid solution of collagen was prepared and the GF was added. The solution was added to the scaffold and it was kept at 4°C for 4 h. It was then vacuum-dried. Colonization was done from a chondrocyte suspension. A micrograph of the colonized scaffold follows (Figure 4.34).

Immobilization to Polyvinyl Chloride

Plasticized PVC has exceptional chemical and physical properties providing an ideal support for direct immobilization of enzyme. Various enzymes have been immobilized on PVC. Covalent immobilization of enzymes onto surface of a PVC sheet is one of the widely available, economic, chemically inert, corrosion-free, tough, lightweight, and maintenance-free. PVC provides a high strength-to-weight ratio.

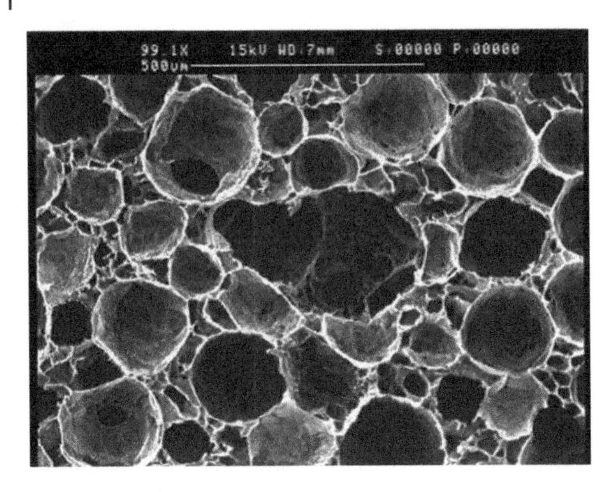

Figure 4.34 Colonized MMA-graphed PLLA scaffold.

A method for attachment of a biomolecule is by what is referred to as a spacer molecule. That is to say that the PVC must first be "activated" to produce an NH_2 or OOH site for binding. This pretreatment you will recall is not unlike that used for PE. The reason for using a spacer molecule is for steric reasons.

Glutaraldehyde is used for this purpose. It possesses unique characteristics that render it one of the most effective protein cross-linking and immobilization reagents. It can be present in at least 13 different forms from monomer to polymer depending on solution conditions such as pH, concentration, temperature, and so on. Accordingly, glutaraldehyde reacts with proteins and substrates by several mechanisms [43].

It reacts rapidly with amine groups at around neutral pH. Studies of collagen cross-linking reactions with monoaldehyde and dialdehydes having chain lengths of two to six carbon atoms demonstrated that the reactivity in this series is maximized at five carbons; thus glutaraldehyde is considered the most effective cross-linking agent [44].

Activation of Polyvinyl Chloride

The inactivity of PVC, while an advantage in use, prevents it from reaction with a glutaraldehyde. Thus with the other olefinic polymers mentioned earlier, it must be activated. Once activated, it reacts with glutaraldehyde to develop the proper linkage for binding to biomolecules.

Activation of the PVC is particularly harsh. It is incubated in a 5 : 1 concentrated nitric and sulfuric acid solution for 24 h. This degrades the polymer surface and produces end groups. The chlorine is removed from the end groups and double bonds are created. It is these end groups that are the sites that react with the glutaraldehyde [45]. This last step is accomplished after thoroughly washing the activated PVC and then incubating it with the glutaraldehyde. The material is then ready for immobilization of biomolecules.

Oxalate oxidase was immobilized on the activated PVC surface. The scanning electron micrographs showed the microstructures on the PVC sheet surface, revealing the successful immobilization of oxalate oxidase. The immobilized enzyme retained 65% of specific activity of free enzyme [46].

Ceramics

Ceramic scaffolds, such as hydroxyapatite and tricalcium phosphate (TCP), are used for bone regeneration applications. Included in this category are various glasses. These scaffolds have high mechanical stiffness, little or no elasticity, and a hard surface. When applied to bone applications, they exhibit excellent biocompatibility due to their chemical and potentially structural similarity to the mineral phase of native bone. In addition, osteogenic cells colonizing ceramic scaffolds are known to be enhanced through osteoblast differentiation and proliferation. Various ceramics have been used in dental applications. Other clinical applications for tissue engineering have been limited because of their brittleness and difficulty of shaping for implantation. Hydroxyapatite is a primary constituent of bone and thus would seem to be ideal as a scaffold for bone.

Immobilization to a ceramic scaffold is problematic but often done.

Summary

In the last two chapters, we have presented the case for movement away from separating the immobilization of biologicals with the associated benefits and the concept of a scaffold with appropriate properties, not the least of which is flow-through, high surface area, and biocompatibility. Progress in this has been inhibited by the skill sets of the individuals. Enzymes and cells are clearly biological disciplines, while scaffolds are the realm of the polymer chemist. We have had the opportunity to have applied our chemistry to the development of a number of medical and environmental products. Our research and production activities have led to a change in the ways chronic wounds are treated in the world. I don't think it is an exaggeration to say that the work of my predecessors and my lab has positioned hydrophilic PU as the preferred dressing for dermal ulcers. Our work in the environmental applications is less developed but in our opinion holds great promise. It was through this research and a passion for the possibilities that led us deeper into applying the uniqueness of PU chemistry to more and more specialty applications.

As you see from the discussions, we are supported by the help of farsighted researchers around the world who have shared our vision. We hope that this book expands that number.

We wanted to pay special attention to the covalent immobilization to PUs. The use of these polymers would appear to put off researchers the opportunity to participate in the field of immobilization. Whether through the use of commercial prepolymers or to the development of unique prepolymer systems, the chemistry and procedures are clear and accessible. Thus it is appropriate that we use a section of this chapter on this chemistry.

Polyurethane Immobilization

In the last chapter we covered the immobilization of biomolecules by several methods. We applied those techniques to a number of substrata. We discussed the immobilization using PUs by several of those methods but have delayed the discussion of covalent techniques. While PU showed some physical advantages, there was no compelling chemical reason to choose the material over any of the others. It is our opinion that reticulated PU is a near ideal architecture. Mass transport, high internal surface area, interconnected cells, and solvent resistance are some of the aspects that support its use as a scaffold for tissue engineering and environmental remediation. As we discussed several times, combining some of the immobilization techniques of the last chapter with a reticulated structure would appear to have advantages, especially if the object of your research involved passing a fluid through it.

The purpose of this chapter is to illustrate the efficiency of using PU chemistry for covalent immobilization. We will review some of the chemistries discussed earlier and then compare it with how the same reaction would be done using isocyanate systems.

Fundamental Principles

We will quickly review the chemistry that makes this possible, but it is appropriate that we begin with the research on immobilizing enzymes using the precursor of the hydrophilic PU prepolymers as we know them today.

A little background might be interesting. Researchers at the Colombia Research Division of WR Grace & Co. investigated a PE-based polyol and a diisocyanate as a base for a flexible and water-soluble offset printing plate. The goal was a prepolymer that would be made photosensitive by derivatizing it with an acrylate at the end groups of the prepolymer. A sample of the prepolymer was accidently left uncovered on the bench overnight. It reacted with the humidity in the air and produced stable foam. This was the beginning of the product that came to be known as Hypol hydrophilic PU prepolymer, currently sold by Dow Chemical. Applications range from agriculture to chronic wound dressings. Currently, Hypol-like materials, based on PEG and toluene diisocyanate (TDI), methylene bis-diisocyanate (MDI), or other isocyanates are manufactured and sold by several companies in the United States and Europe. Foams made from those prepolymers are also available.

During the development of Hypol, L. L. Wood in the Grace lab investigated the reaction of isocyanates with biomolecules, specifically enzymes. This research led to literally hundreds of projects around the world. A search of key words “Hypol” and “enzyme” confirms this.

Having said that, and in the context that researchers should not be bound to commercial versions of PUs, we will look at the Wood patent (USP 4,312,946) in some detail. It begins with the development of a prepolymer, and so it will help relieve the researcher to develop custom materials. As we have already discussed in Chapter 1, there are a number of variables within the structure of the prepolymer. Other than the isocyanate and the polyol, molecular weight of the polyol and the amount and type of cross-linker are all variables that the researcher can use to build a prepolymer that best meets the design of the ultimate product. The examples in the Wood patent give us sufficient information to be able to customize a prepolymer.

Referring to the Wood patent, in the first example, the basic process to make a hydrophilic PU is described (Table 4.7). A prepolymer is prepared in which 690 g of PEG of molecular weight 1000 and 310 g of pentaerythritol was reacted with 1830 g of TDI. Pentaerythritol is the cross-linking agent and improves heat stability. In addition, you will remember that a cross-linking system is needed if foam is to be the product. Trimethylolpropane works well and is probably the triol of choice for commercial production.

The mole ratio (TDI to polyol) in this experiment is 2:1. The reaction takes a few hours to react at about 65°C. In laboratory practice, the viscosity is monitored. The viscosity increases rapidly as the reaction comes to completion. The product of the reaction is a classic prepolymer.

In the example, the PU prepolymer is cooled to 4°C and a solution containing trypsin is added. The reaction is exothermic and has the potential to denature the enzyme. The mixture was stirred until it began to foam. After foam, it was washed in the buffer solution. The activity of the trypsin was confirmed by placing it in a 1% casein solution.

Table 4.7 List of Ingredience in example 1.

	Molecular weight	Equivalent weight	Mass	Equivalents
Polyethylene glycol	1000	500	690	1.4
Pentaerythritol	136	34	310	9.1
Toluene diisocyanate	174	87	1830	21.0

From Wood [47].

As an aside, and not mentioned in the patent, was the quality of the foam. The cell structure of a PU is strongly affected by the temperature of the reacting mass. At the low temperature used in the first example, the result would have been a closed-cell foam. In order to do the activity, one would have had to cut the foam into small pieces to increase the surface area.

In a second example we learn something about these limitations of the Wood technique. A 10% solution of pectinase in a phosphate buffer was mixed with the prepolymer. The resultant foam was *cut into pieces* and placed into a column. Apple juice was passed through the column, which clarified the apple juice.

Stepping away from Wood for a moment, in the chapter on laboratory practice, we discussed the use of surfactants to assist in the emulsification of the prepolymer. Depending on the surfactant, one can make ultrafine core structure of uniform size or large and nonuniform pores. In either case they are considered "open cell." By comparison, without a surfactant, a closed-cell foam develops. As we said earlier, the cell structure in part is determined by the characteristics of the aqueous/prepolymer emulsion. Without an emulsion aid, closed cells are inevitable. While that may be the desired architecture for some applications, it is not usually what is wanted. We recall a product a number of years ago in which the emulsion temperature was so high that after a few seconds, the foam collapsed to form a closed-cell structure. It was all but useless except for the product it was used for. Typically, however, an open structure will be the goal. That does not mean that the foam can be considered a flow-through device, however. Two factors need to be considered in the development of a true flow-through device. While open cells are necessary, the term does not sufficiently describe the structure. The following micrograph of an open-cell foam shows that while a liquid will flow through the foam, it would be what is termed a tortuous path (i.e., not straight through) (Figure 4.35). The inability to pass liquids through the foam is hampered by the structure. The difference is in the interconnectedness of the cells. It has become common in the relevant literature to use the terms porosity or void volume to describe foam intended for use as a scaffold. The foam in the figure has a high porosity and void volume, but those properties are not sufficient for our purpose. The diminished interconnectedness would result in high pressure drops and inhibit cell proliferation.

Another example is useful. Hydrophilic foams are often used in the treatment of chronic wounds. One of the best in our opinion was developed by Rynel Corp. in Boothbay, ME, USA. It was made for Johnson & Johnson of Arlington, Texas, under the trade name SOF-FOAM® wound dressing. It was an open-cell foam of high porosity and void volume, but using careful temperature and formulations, we were able to build the foam such that it would hold from 15 to 20 times its weight in water. This is not uncommon, but unlike other dressings, the process minimized the

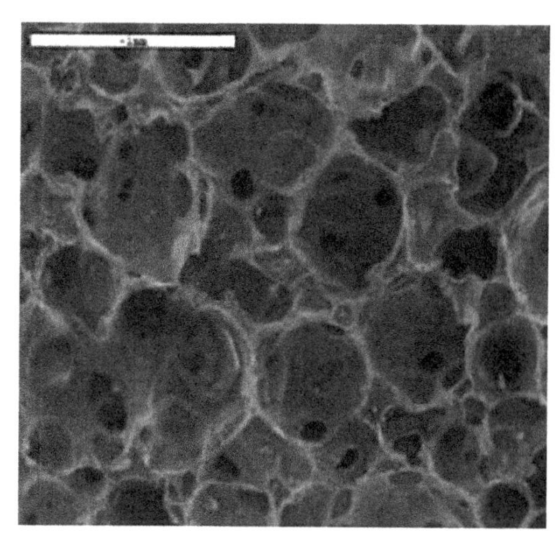

Figure 4.35 Open-cell foam.

level of interconnectedness, thus ensuring that most of the water would not drain when held up by a corner. Thus while it had a high void volume, it had practically no flow-through capability.

In immobilization, high interconnectedness is our goal. For this purpose we have found it useful to adapt a method taught in a subsection of ASTM standard D3574 that measures the flow it takes to flow air through foam. We discussed this in the laboratory chapter. The method has two parts. One part keeps the pressure the same and measures flow rate (the method we prefer), and the other measures the pressure needed to achieve a certain flow rate. Consult the standard for the procedures.

We had mentioned two factors with hydrophilic foams, which because of its low compressive strength present another issue. The pressure drop across the foam is typically higher than the compressive strength of the foam, and so it will begin to collapse. Accordingly the pressure drop increases, eventually blinding off the foam altogether.

The Wood patent went on to describe other aspects of the immobilization phenomenon. In a third example, the prepolymer described earlier was used to immobilize urase. To quote the example, "After 4 months of continuous operation, the column was still active in the hydrolysis of urea to ammonia and carbon dioxide as indicated by a pH rise in the effluent solution."

To expand the scope of his research, he made a prepolymer using PPG. Details are lacking, but for completeness, we include it. The PPG was mixed with an equal mass of TDI. The mixture was heated and stirred until a homogeneous solution was achieved. The completeness of the reaction was not described, but experiment teaches us that a reaction occurred. After cooling, a solution of amyloglucosidase was added. The material eventually developed into foam. The product was "crystalline" PUF. It was shown to be active in hydrolyzing a starch solution.

This work formed the basis for the interest in PU immobilization. It is now our task to expand on the subject in the context of biochemical progress made since 1973.

We must make an important comment. If you learn one thing from this book, it is that while immobilization is a necessary step in the development of tissue (not simply the propagation of cells), it is not sufficient. Coupling this with an appropriate scaffold is the gateway to continued progress. It is, in our opinion, that continued studies on flat plate systems are only a prelude for constructive but difficult research. If one believes in the evolution of life, one must admit the necessary function of scaffold. To paraphrase the Churchill quote, eating chicken cells is not the same as chicken wing or breast. The design of an appropriate scaffold includes architecture and chemistry. We've covered architecture, so we have turned our attention to the chemistry pioneered by Wood.

Prepolymer Chemistry

As we described earlier and in the chapter on chemistry, a prepolymer is produced from the reaction of a diisocyanate and a polyol. The polyol could be a PEG for a hydrophilic or a PP or a polyester for a hydrophobic prepolymer (Figure 4.36).

Prepolymers are liquid intermediates between monomers and a final polymer. They are formed by reacting an excess of either polyol or isocyanate so that the product is still liquid and contains the reactive functional group of the reagent in excess.

Almost important, it allows the medical or environmental researcher to explore this immobilization technology without the need of a chemist. Prepolymers are available at moderate cost. The supplier can give the researcher with a certificate of analysis or compliance. This allows the material to be used "without purification."

Still further, as we will discuss, it is a polymeric system, the surface of which is "activated" and therefore ready for immobilization. Compare this with the activation techniques for PE earlier.

Lastly, while a fume hood is recommended, the vapor pressure of the diisocyanate is reduced in prepolymer formulations. This is especially important

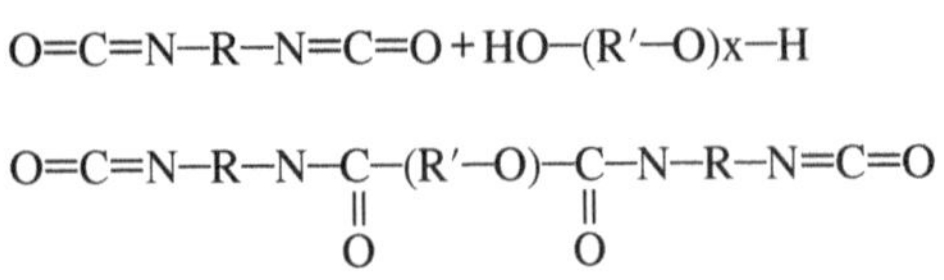

$$O{=}C{=}N{-}R{-}N{=}C{=}O + HO{-}(R'{-}O)x{-}H$$

$$O{=}C{=}N{-}R{-}N{-}\underset{\underset{O}{\|}}{C}{-}(R'{-}O){-}\underset{\underset{O}{\|}}{C}{-}N{-}R{-}N{=}C{=}O$$

Figure 4.36 The prepolymer reaction.

when a TDI-based foam is desired. It is important to mention that TDI has a sufficiently high vapor pressure to be toxic, and so a fume hood is required for work with the isocyanate. In addition, a small fraction of the population has a strong allergic reaction to TDI, and so they should be restricted from exposure.

The reaction shown earlier is generic in the sense that it could be hydrophilic or hydrophobic. The difference of course is the polyol. Both are commercially available. The latter is typically sold in two parts. Part A is the prepolymer and Part B is a polyol, surfactants, and catalysts. Mixing the two produces either an elastomer or a foam. Hydrophilic prepolymers are sold, but without a "Part B." The aqueous phase is prepared by the researcher. The technique is described in the laboratory practices chapter. For the purposes of this here, however, the biomolecule is typically dissolved in the aqueous.

While it is possible to immobilize biomolecules to a hydrophobic PU, most of the work in this field is done with hydrophilics. The physical nature of the process dictates this. As you will see, the molecule is typically dissolved or at least dispersed in the aqueous phase and then emulsified with the hydrophilic prepolymer. With that in mind, while we suggest that your research begin with a commercial prepolymer, the range of formulation while still maintaining hydrophilicities is very broad. The degrees of freedom include a wide variety of isocyanates. Of particular interest might be isophorone diisocyanate for toxicity reasons. While MDI and TDI are aromatic and subject to yellowing, one might chose aliphatics like hydrogenated MDI. With regard to the polyol, molecular weight is an important variable. From the chemistry chapter you will remember that the polyol is a "soft segment." Increasing or decreasing the molecular weight dramatically changes the texture of the foam.

We have discussed the cross-linking phenomenon at length and so it won't be repeated here. What we have not discussed are chain-terminating additive. As the polymerization proceeds, it goes through a range in which it is adhesive.

If you are new to this technology or don't have the option of doing your own chemistry, we would strongly suggest that you use a commercial hydrophilic prepolymer like a Hypol prepolymer. You literally can efficiently immobilize a biomolecule within a few minutes of receiving the prepolymer.

The Immobilization Chemistry

By way of review there are two reactions that when combined permits the immobilization. The first of those is necessary because it begins the polymerization process. We refer to it as the water reaction. It goes without saying that this step generates the internal pressure that produces foam (Figure 4.37).

$$R{-}N{=}C{=}O + HOH \longrightarrow R{-}NH_2 + CO_2\uparrow$$
(The isocyanate) (Water) (An amine)

Figure 4.37 The water reaction.

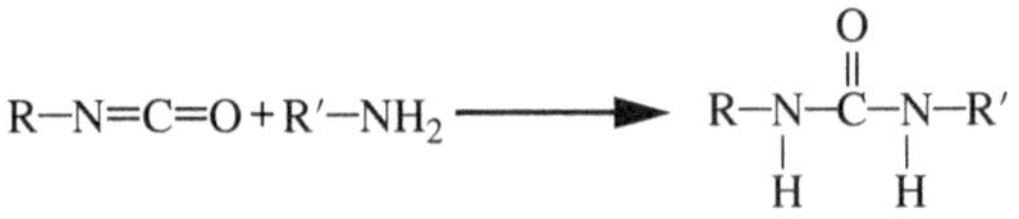

$$R{-}N{=}C{=}O + R'{-}NH_2 \longrightarrow R{-}\underset{H}{N}{-}\overset{O}{\overset{\|}{C}}{-}\underset{H}{N}{-}R'$$

Figure 4.38 The amine reaction.

It is important to remember however that this gas would simple bubble away if it weren't for the coproduct of the reaction, an amine. As an aside, one of the few molecules that do not react with an isocyanate is another isocyanate. This makes PU prepolymers possible.

To continue the review of chemistry, you will also remember that the amine product of the water reaction reacts with an isocyanate to produce a urea linkage. This is the primary polymerization reaction. If the prepolymer included a cross-linker, the molecular weight increase also starts to build a gel structure. It is this gel structure that builds the foam. As the strength of the structure increases, despite the internal pressure developed by CO_2, the foam resists, increasing in volume. The pressures are released by breaking the windows between the pores. The result is a relatively low density open-cell foam (Figure 4.38).

This is the natural course of the PU reaction; it is also the dominant reaction used to covalently immobilize molecules into a PUF.

Structure and Chemistry of Biomolecules

For review, enzymes are linear chains of amino acids. The sequence of the amino acids defines the structure and the catalytic activity of the enzyme. The activity of an enzyme is, in part, defined by structure. The activity also involves the sequencing of the amino acids. The structure is typically described as "globular." Enzyme like all complex proteins denature at moderate temperatures, typically around 40°C. I can express the opinion that this process is not a change in the protein directly but rather a change in the property of the water in which the protein exists. Water has a definable structure that changes with temperature and ionic strength.

The target of an enzyme is referred to as the scaffold. We mentioned earlier that we prefer to use the term scaffold for the support structure that we use for immobilization. Grammatically, we could use substrate as well, but you can imagine the confusion. So to be clear scaffold is for architecture and substrate is for enzyme activity. The size of enzymes ranges from just 62 amino acid residues [48] to over 2500 residues in some fatty acid synthases [49]. Despite their size a small portion of their structure (around 2–4 amino acids) is directly involved in catalysis [50]. This catalytic site is located next to one or more sites

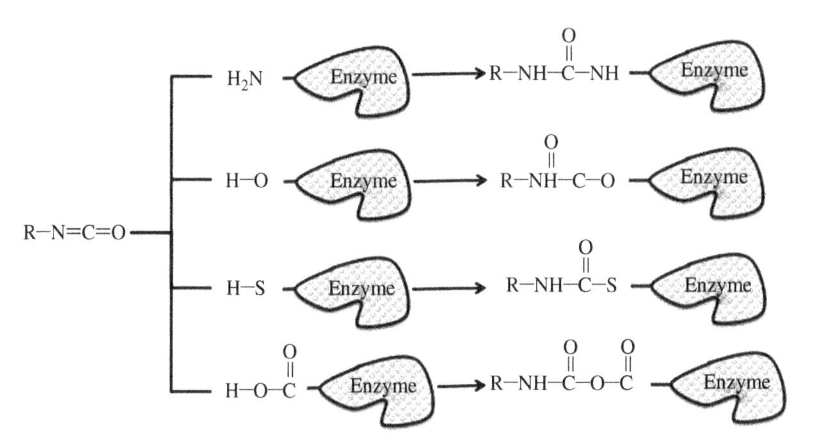

Figure 4.39 Reactions of isocyanates (R—N=C=O) with enzyme surfaces.

that act as binders to the substrate. The binding site plays the critical role in orienting the enzymatically active site of the protein. You will remember from basic chemistry that reactivity, that is, how two reactants approach one another, is a contributing factor in reaction kinetics. The binding mechanism is the reason enzymes increase both the rate and the specificity of the reaction.

Immobilization of enzyme using a PU is based on the reaction of various components on the enzyme surface. While this could be the binding site or the site at which the catalytic reaction takes place, statistic favors the noninvolved site. The Figure 4.39 to follow lists the sites that could react with the isocyanate [51].

Preparation of Immobilized Biomolecules

We have tried, perhaps unsuccessfully, to broaden our discussion beyond the use of hydrophilic PU prepolymers to immobilize biomolecules. The ease at which the immobilization can be done with a commercial prepolymer unfairly prejudices the discussion. It is significant that non-chemists can do innovative research on immobilization, biocompatibility, cell colonization, and so on, minimizing the full impact that PUs can have. That impact begins with the physical properties including tensile and compressive strength and hydrolytic stability. As we have stressed, the architecture of a reticulated PU has been shown to have great potential. It is a subgoal of this text to combine these properties into outcome-based devices. As we will show in the examples that follow, limiting the research to commercial hydrophilic prepolymers is neither necessary nor appropriate. Having said that, if you are new to the technology, start with a commercial product.

In those examples that involve a hydrophilic prepolymer, little attention is made to the structure of the foam. This is unfortunate and diminishes the potential. The default technique is, as we have often mentioned, to cut the foam into pieces. In our opinion this is only slightly more appropriate than casting a film. This is especially unfortunate in the context that the control of the architecture is straightforward. Not that one could mimic a reticulated foam but with a little preparation the foam could develop some mass transport properties that would enhance activity. The foams would doubtlessly still be cut into pieces, but the increase in activity should be measurable. When you approach this technology, it is best to realize that care must be taken inasmuch as at least two competing reactions are taking place: the enzyme–isocyanate immobilization and the water–isocyanate reactions. The water reaction and subsequent amine chain extension build the foam structure, physical strength, and architecture, while the enzyme concentration determines the final activity. To optimize the results, both concentrations and temperature must be controlled. Put another way, your prepolymer has a given amount of isocyanate functionality. It is split between unreacted diisocyanate and the end caps of the polyol. The diisocyanate is most reactive, but the enzyme that reacts with it is most probably water soluble and would be washed out. If you are using a commercial prepolymer, prereacting the prepolymer with a monoamine might be advisable. Then one can be confident that the enzyme would be bound to the larger polyol molecules.

We will focus on a number of illustrative examples of researchers whose focus was on the immobilization process. We are not limited in this review, however, on project specific to hydrophilic prepolymers. We have tried to make the point that PUs are somewhat unique in that without complicated chemistries, the researcher can build unique molecules. A paper by researchers at Rice University referenced later, rather than using a PEG, used a polypeptide to increase endothelialization. This study was done on flat plates, but one would have to wonder how this could be adapted to a true 3D scaffold. Our goal is to develop a biologically active surface within an accessible structure. Critical to progress in that arena is the research on immobilization without regard to scaffold.

Thus studies of immobilization are very useful and demonstrate the efficacy and ease to which the technique is performed, particularly with prepolymers. Future research should include a nod toward how the properties of the immobilized devices will be used *in vivo* or at least extracorporeal. We have mentioned it before but it deserves repeating. If we are to go through the trouble to make foam, we should, as a matter of practice, ensure that the internal surface is available.

If a foam is the goal of immobilization research, it is our recommendation that researchers consult the manufacturers literature for instructions on how to control the cell structure of a foam outside the context of immobilization.

Still further, for those whose work is on 3D architectures, use the techniques discussed earlier to fully describe the quality of the device. As we said, porosity and void volume are not sufficient.

To use an environmental example, the tobacco processor we discussed earlier used reticulated PUF as a scaffold to support an active biofilm. The engineering design required it to process a large volume of air quickly. In order to accommodate that, the engineering firm cut the foam into 40 mm cubes and loaded it into a vessel. Realizing that unless accommodations were made, the air would flow over the cubes, but not penetrate the inner structures. Therefore foams in the range of 20 ppi were used, making the internal surfaces available for remediation.

Staying with foam structure before we discuss some examples from our research and that of others, our most important tool in controlling foam architecture is by the emulsion that we create with the prepolymer and the aqueous phase. Review the chapters on chemistry and laboratory practices for details. In those instructions you will see examples of foam structure (openness and pore size) as affected by surfactants. As a general rule, the pores that are created are a function of the quality of the emulsion that is produced during mechanical mixing. Pluronic L62 (BASF Corp.) is the standard for a fine cell foam, but not as a flow-through architecture. Typically it is used at about 0.1% by weight based on the water phase. Many wound dressing formulations are based on this recipe. Its mass transport characteristics, however, are improved by a small amount of Pluronic F88. This is an example, however, and even better formulations are possible given the degree to which the researcher wants a device to perform. Continuing examples of foam control, if a stiffer foam is desired, the investigator might consider an MDI-based prepolymer.

While formulation is the starting point, the temperature of the curing is critical to a successful foam. Remember that two reactions are taking place, each of which has unique activation energies and, therefore, kinetics. In general, all things considered, lower temperatures favor the polymerization reaction. This results in denser, more closed cells. High temperatures favor the liberation of CO_2, which can produce low density and weak foams but in the extreme can result in a collapsed foam.

Simultaneously, CO_2 is being released while strength is building through polymerization. If polymerization is favored, the internal pressure caused by the evolving CO_2 will certainly break through but not until the polymer has developed too much strength to expand to its fullest extent. At high temperatures, CO_2 breaks down the structure of the foam before it can develop an organized architecture.

One must remember that these reactions are exothermic and this must be considered. Internal temperatures of greater than 70°C are not uncommon in the manufacture of any PUF. The temperature in the center of a large block approaches the adiabatic. This, of course, will denature most common enzymes and you see this in the examples to follow.

We have selected a number of examples of innovative uses of PUs for immobilization. In some cases the immobilizations have been done on surfaces but are also developed foams, typically without regard to the interconnectedness. The list includes several that we sponsored and/or participated. In those cases, a prepolymer was grafted onto a reticulated foam.

While hydrophilic prepolymers are well represented in this catalog, other chemistries are described. Because of the frequency of hydrophilics, however, we suggest that it would be useful to review the typical methods used to immobilized biomolecules. These start with commercially available prepolymers, but following that, options are available to meet individual product goals. As we said, you don't need a chemist. With proper ventilation, you can do this at home. Mixing water with the prepolymer is the most common technique. Several of the papers are based on a prepolymer being mixed with water containing a surfactant and a biomolecule. While not mentioned specifically, if you intend to make a film, you can dissolve the prepolymer in acetone. By this technique thin films are possible. It is also useful as a way to coat inorganic or organic preformed scaffolds. The result is a scaffold coated with a hydrophilic surface to which a biomolecule is immobilized. With that, consider the following research.

Notable Uses of Polyurethane for Immobilization

We want to recognize that you in your personal research are not bound to commercial PUs, although it is an appropriate start. The only component that is required to be called PU is the use of a diisocyanate. Given the breadth of isocyanate molecules, this gives the researcher a broad spectrum of possibilities.

A case in point is work done at Rice University that we described earlier that moves from simply making the polymer compatible to partially mimic the *in vivo* environment by inserting a polypeptide sequence into the PU backbone. One of the aspects of this text is that the information herein allows those not expert in chemistry to participate in these advanced synthesis exercises.

Organophosphates

While the mass transport of fluids through a medical device is debatable and application specific, low pressure drop systems are a requirement of environmental remediation. High pressure drop scaffolds affect the cost of operation and capital equipment in a market where cost control is critical. High on the list of applications would be an effective system to remediate contaminants in groundwater runoff. The amount of pesticides, herbicides, and so on used in the United States makes contamination of groundwater inevitable (Table 4.8).

Organophosphates (OPs) are a class of insecticides. Thirty-six of them are presently registered for use in the United States, and all can potentially cause acute and subacute toxicity. OPs are used in agriculture, homes, gardens, and veterinary practices. In the past decades, several notable OPs have been discontinued for use, including parathion, which is no longer registered for any use, and chlorpyrifos, which is no longer registered for home use.

Due to the extensive use of this class of pesticides, it is inevitable that significant quantities will find its way to the water supply system (Table 4.9). The persistence of OPs in water at 10°C was reported by Muhlmann *et al.* [52].

For the removal or reduction of these water pollutants to acceptable levels with membranes, steam distillation, oxidation, and adsorption on activated charcoal have been reported. These pre-concentration methods are often slow or cumbersome or are too expensive.

Table 4.8 US pesticide market by function and volume, 2007.

	US market	
Type	**Mil lbs**	**%**
Herbicides	531	47
Insecticides	93	8
Fungicides	70	6
Other	439	39
Total	1133	100

Source: Adapted from Mülhmann and Schrader [52].

Table 4.9 Persistence of organophosphate pesticides.

Compound	Half-life in days at 10°C
Paraoxon	1200
Parathion	3000
Dipterex	2400
Methyl parathion	760

Source: Adapted from Mülhmann and Schrader [52].

A team of scientists at the Commonwealth Scientific and Industrial Research Organization (CSIRO) and Orica Australia were working to find an efficient, cost-effective method to clean up pesticide residues. Preliminary research had isolated enzymes that will degrade OPs and other insecticides. The research started by looking for enzymes in a group of bacteria found in the soil. These naturally occurring bacteria feed on chemicals such as insecticides and can break them down into nontoxic compounds. Bacteria were found that were able to break down some classes of pesticides and then isolated the enzymes responsible. After the genes responsible for producing the enzyme had been identified, they were then cloned into a common bacterium such as *Escherichia coli*. This makes it easier to study them and allows the production of large quantities of bacteria in industrial fermenters. The enzymes were then extracted.

Orica Corp. (East Melbourne Victoria, Australia) carried out a field trial on a cotton farm in 2001. The results showed that the enzyme reduced residue levels in more than 80 000 liters of contaminated drainage water by 90 % in only 10 min.

Two problems, however, were recognized: The first was the half-life of the enzyme. The second was not recognized at the time but developed as our phase of the project developed. It was their goal to develop a skid-mounted system that could be moved from site to site to remediate ponds and tanks. Accordingly, a flow-through system of immobilized enzymes was anticipated.

After reviewing our patents we were contacted and entered into a contract to develop a flow-through immobilization system using their enzyme. The result was a confidential report, but we can review some of the salient points.

We used a 30 ppi reticulated foam supplied by FXI Corp. (Media, PA, USA). It was coated with an emulsion of Hypol 2002 and an aqueous solution of 0.05% Pluronic L62 and the enzyme. The emulsion was applied to sheets of foam and passed through nip rollers. It was then flipped over and passed through the rollers, again assuring uniform coating. After curing (about 1 h), it was sealed in a container. Experiments were conducted on the wet foam.

The results of the experiments are confidential, but suffice it to say that the enzyme was active at degrading an OP (supplied by Orica) and remained so for 7 days without noticeable reduction in activity (the length of the study).

During the investigation, however, it was noticed that the solution changed color. It was decided that the phosphate salt was itself considered a pollutant and needed to be removed. You will remember the discussion of extraction from the last chapter that PUs are an effective medium in this regard. The solution was passed through a hydrophilic PU composites and the phosphate was removed. No claim, however, was made that this was an effective or economical method.

Lastly, as we said, the object of the study was to develop data that would be used to design a skid-mounted process. To supplement the study we assembled a column of the hydrophilic-coated reticulated foam (30 ppi) and packed it into a 1 m long, 10 cm diameter PVC pipe with appropriate fittings. The flow and pressure drop was shown to be sufficient for the project.

Lipases

Lipases are an important group of biocatalysts. They have high catalytic activity, but as is typical of enzymes their half-life is short. As with other enzymes, immobilization typically decreases the catalytic activity but improves stability. This offers the advantage of lowering operational costs. Lipase immobilization allows for better control, easier product recovery, flexibility of reactor design in some cases, enhanced storage and operational, and thermal and conformational stability. The activity and stability, however, depend in part on the type of support and immobilization method.

In a recent report [53], two hydrophilic PU prepolymers were investigated as scaffolds for lipase immobilization: Hypol 2002 is a TDI-based prepolymer and Hypol 5000 is based on MDI. Both use PEG 1000 as the polyol. Hydrophilic PUFs were prepared by mixing the PU prepolymer with the aqueous phosphate buffer solution containing lipase enzyme.

A continuous packed-bed reactor was used to test for the esterification at 30°C. Both TDI and MDI foams were used. The foams were cut in cuboids (~0.07 cm^3). The solution was continuously pumped upward at a flow rate of 0.1 m/min for a residence time of about 260 min. Effluent samples were taken and assayed for ethyl butyrate and ethanol.

The stability of the lipase-immobilized PUs was tested in batch mode as well. Both biocatalysts showed a high operational stability along the continuous 30-day operation of a packed-bed reactor, when reaction media with low substrate concentrations were used. Conversely, under high substrate concentrations, a fast deactivation of the biocatalysts was observed in continuous batch mode. The low operational performance was thought to be the inhibitory effect of ethanol, which tends to accumulate inside the foams.

In another study [54] in which we participated, a lipase enzyme solution was emulsified with an equal volume of an MDI-based hydrophilic PU prepolymer (Urepol 1002) using a MIXPAC System 50 handheld meter–mix–dispense device. The liquids were mixed at room temperature. The emulsion was immediately applied to 12 pores/cm reticulated PUF and passed through a set of pinch rollers in such a way as to coat the entire inside structure of the foam (Table 4.10). The foam was tack-free in 2 min and fully cured in 1 h, after which it was air-dried at ambient temperature.

In addition to the enzymatic activity, the paper presented a schedule for describing the scaffold on which the immobilized enzyme resides. Specifically, data on flow-through characteristics of the scaffold is presented. It was beyond the scope to measure surface area, but it could be estimated from the reticulated PU before coating. (Again we partly funded and participated in the development of this research.)

The researchers concluded the study by summarizing the project. The foams were found to be active, but somewhat less than a commercial

Table 4.10 Foam for lipase study.

Property	Value
Pores (cm)	18
Density (kg/m^3)	30.4
Deflection (g/cm^2)	36
Tensile strength (kg/cm^2)	1.1
Elongation (%)	180
Tear strength (kPa)	27.6
50% deflection (kg·cm^2)	1.07

Source: Vasudevan *et al.* [54]. Reproduced with permission of Springer.

lipase-immobilized resin. Nevertheless, the immobilization technique was determined to be effective and simple. In addition, the low resistance to fluid flow appeared to make it a viable support to be used in large reactors. The enzyme could easily and quickly be loaded in any packed-bed reactor.

Fibroblasts

We will deviate from the main theme of this chapter and discuss some cell adhesion studies that we sponsored at the Maine Medical Center in Scarborough, ME, USA. We wanted to investigate the attachment of cells to the hydrogel-coated reticulated foam in US patent 6177419. While these studies are not definitive of either the material or the adhesion mechanism, they do offer some optimism that the material has potential to become an effective scaffold for tissue development.

Accordingly we sponsored some preliminary cell adhesion studies at the Maine Medical Center Research Institute (MMCRI) under the direction of Dr Igor Prudovsky. They were to study the attachment of NIH/3T3 mouse fibroblast and other cells to a composite material supplied by us.

The material was made by dissolving Hypol 2002 in acetone (10%). A 30 ppi reticulated foam was immersed in the solution. After a few minutes it was removed, allowed to drain, and then dried at 50°C for 2 h. The material was placed in a water with 0.05% Pluronic L62 to cure the prepolymer coating. They were dried and packaged for delivery to MMCRI. Analysis showed that the composite had a coating of 11.5% by weight.

In the first of several studies, the material was treated by MMCRI technicians with fibronectin and then inoculated with the NIH/3T3 mouse fibroblast for 48 h. The following micrograph is exemplary of the result (Figure 4.40).

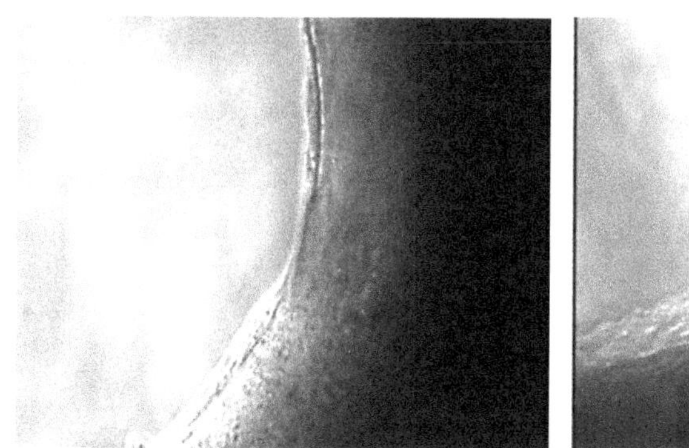

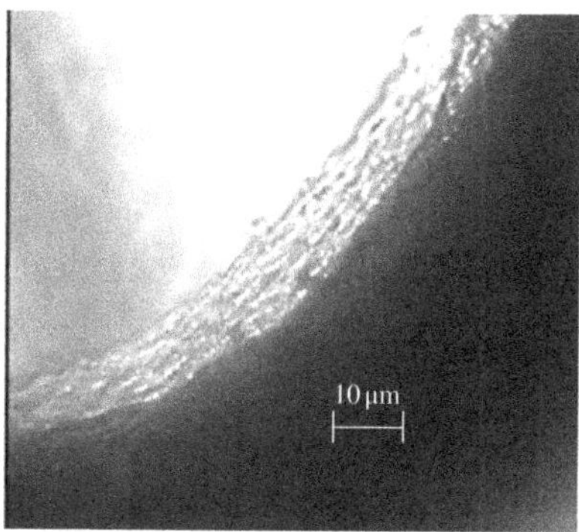

Figure 4.40 NIH/3T3 cell attached to the fibronectin-imbibed scaffold (left 48 h right 96 h) (unpublished data).

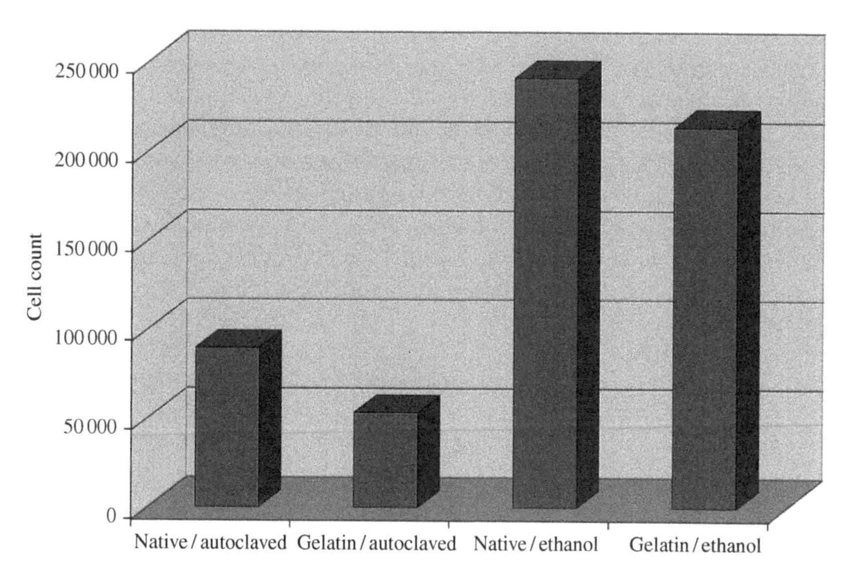

Figure 4.41 Cell growth of smooth muscles by composite scaffold by pretreatment. (*See insert for color representation of the figure.*)

Smooth muscles (PAC) were also evaluated for growth and attachment using the same techniques as NIH/3T3 cells (Figure 4.41). With this cell type, the chemistry of the substratum did not seem to affect the ability of the cells to attach. The highest cell numbers were observed in the untreated scaffold sterilized with ethanol.

Endothelial cells on the fibronectin-imbibed composite were also propagated with good success.

Based on these studies, Dr Prudovsky concluded:

> The advantage of the composite over other polymeric materials currently used as substrates for three-dimensional (3D) cell growth is not only the unique combination of hydrophilicity, elasticity, easy formation of structures with controllable pore size, but also the potential to regulate substrate degradation and introduce its modifications, which enhance or decrease cell adhesion. This combination of characteristics makes HPUFC (ed. the composite) a highly promising substrate.

Collagen

Polymer surfaces are commonly used for biomedical investigations. Typically however, surface energies such as hydrophobicity interactions interfere with biological molecules at the hybrid interface. If the polymer is biodegradable, degradation generally begins at the surface. This change in the surface chemistry can be cytotoxic. Vegetable oils have a number of excellent properties, including low toxicity and high biodegradability. In a recent paper [55], undecenoic and oleic acids were derivatized into diols and were then reacted with MDI to produce a PU prepolymer. According to the authors, PUs are frequently used in the fabrication of scaffolds for tissue engineering as they are available in a wide variety of compositions and properties and can be easily shaped into complex structures. The goal of their study was to produce a vegetable-based PU that enhances cell, blood, or tissue compatibility.

Quoting from their paper,

> For the engineering of hard tissues, the scaffolds should have an adequate structure to maintain the spaces required for cell-in growth and matrix formation during in vitro culturing. Moreover, they must provide sufficient temporary mechanical support, matching the mechanical properties of the host tissue as closely as possible to bear in vivo stresses and loading.

Of course we would add interconnected pore for cell proliferation and mass transport. As you will see, this study uses flat surfaces for their investigation, thus avoiding experimental complications.

Thermosetting PU films were made using the undecenoic and oleic acids using MDI as the isocyanate. An argon plasma was applied to the PU films to activate the surface for grafting of acrylic acid molecules. The concentration of carboxylic groups on the PU surface after the grafting/polymerization treatment was determined calorimetrically.

The films were immersed in an aqueous solution of 1-ethyl-3-(3-dimethylaminopropyl)carbodiimide (EDC) in the presence of *N*-hydroxysuccinimide.

These activated samples were then immersed into a collagen I solution to form stable amide linkages on PU surface. The amount of collagen immobilized on PU surfaces was also determined colorimetrically.

The study confirmed that plasma-assisted acrylic acid monomer polymerization is a useful, if complicated, way to introduce carboxylic groups on fatty acid-based PU surfaces. Collagen was successfully immobilized on the PU–PAA surface. Coupling of collagen molecules had a positive effect on cytocompatibility of PUs. A higher amount of immobilized collagen on PU–PAA surface resulted in increased cytocompatibility.

In another study [56], researchers used a collagen-immobilized PU surface for the controlled release of antibiotics. In this case the surface of a commercial PU was chemically activated. The PU was a hydrophobic aromatic polyether-based thermoplastic PU.

An aqueous grafting solution of acrylic acid was prepared. The PU samples were cleaned, dried, and placed in the grafting solution. Grafting was allowed to continue for 15–20 min at 30°C. The grafted samples were rinsed in DI water and dried at room temperature.

Surface-grafted samples were immersed in a sodium tartrate dihydrate solution adjusted to pH 3.0. The samples were immersed in a 4-morpholinethane-sulfonic acid solution containing 3-ethyl-1-(diaminopropyl)carbodiimide and *N*-hydroxysuccinimide. The carbodiimide activation was performed for 5 min at room temperature. The immobilization was completed by immersing the carbodiimide-activated samples in a buffered solution containing type I collagen.

Discs of immobilized collagen were implanted in male rats (subcutaneously) for up to 6 weeks. It was concluded that the collagen surface modification improved tissue integration. Promising results were reported with regard to reducing the incidence of implant-related infections.

One of the surprising determinations during the preparation of this manuscript was that despite a large volume of research on immobilizing proteins with PU prepolymers, we could only find one reference to their use with collagen. A colleague at Brigham and Women's Hospital in Boston included solubilized type I collagen in a vascular prosthesis made from Hypol 2002 [57]. It was made by mixing an aqueous containing the collagen and a surfactant (Triton X-100) that was reported to yield a closed-cell foam. After mixing, the emulsion was applied to a spinning mandrel to form it into a tube-like structure. While there is no mention of a covalent reaction of the collagen with the isocyanate end groups, it is a logical assumption.

Lastly, the union of cells and polymers has become a common tool for constructing artificial organs for long-term prosthesis. There are two types of biohybrid systems: one that target cells on a biocompatible, nondegradable polymer surface and another that target cells on a biodegradable polymer scaffold. Natural blood vessels are lined by an endothelial cells (EC).

Considerable effort has been put into mimicking this natural vascular system by developing an endothelial cell lining. Cell adhesion can be by adsorption or other mechanisms.

In any case, the adhesion and proliferation of the ECs on a surface are affected by the surface properties. One method of enhancing the adhesion of ECs on the biomaterial is to treat the surface of the biomaterial with collagen. Cells adhere to collagen surfaces that are mediated with integrins.

In a relevant study [58], human umbilical vein endothelial cells (HUVEC) were cultured on a complex interpenetrated network within a PU to investigate how the immobilization of collagen affects EC attachment and growth:

Step One: PEGs with various molecular weights reacted to replace the end groups with acrylate [59]. The PEG was then reacted to form a diamine (PEGDA).

Step Two: A diisocyanate-terminated PU prepolymer was prepared by reacting poly(ε-caprolactone) diol with hexamethylene diisocyanate (HDI) at 65°C for 2 h under a nitrogen atmosphere. The result was diisocyanate-terminated PU prepolymer with a chain extender and a triol to cross-link the polymer.

Step Three: An interpenetrating network was developed by mixing the PU of step two, with the PEGDA of step one. The mixture was cast in a glass plate mold. The network formed when the mixture was exposed to UV light, resulting in a PU–PEGDA polymer. Benzoin was used as a photoinitiator.

Step Four: To prepare the collagen-treated samples, PU–PEGDA-coated Petri dishes were coated with a collagen solution containing acrylic acid and incubated at 37°C for 12 h. Finally, the samples were swollen with a buffer solution for 1 day and rinsed (PU–PEGDA–Coll).

Step Five: HUVEC were seeded into 24-well culture plates that contained collagen-treated PU–PEGDA. The plates were incubated at 37°C under a humidified 5% CO_2 atmosphere.

Note: While the collagen was not described as being covalently bound, the thorough washing and the reaction conditions suggest that covalent bonding was likely.

Conclusion: In any case the researchers successfully cultured hybridized HUVEC on the collagen-immobilized polymer surface. Micrographic examination revealed the successful spread of endothelial cells.

Amyloglucosidase

Amyloglucosidase is an extracellular enzyme that catalyzes the hydrolysis of starch. Amyloglucosidase is used commercially in free form. It is a relatively inexpensive enzyme, but it is accepted that an immobilized system would offer significant advantages in order for industrial utilization to occur.

In the following study [60], amyloglucosidase was immobilized on two hydrophilic PU prepolymers: a foam based on Hypol 2002 and an experimental grade that produces foam. The enzyme was covalently attached to the polymers during the foam or gel process. The combination of the properties of the immobilized enzyme and the inherent properties of the foams/gels was thought to offer advantages.

The foamable prepolymer is a water-activated derivative of TDI, whereas the gelable prepolymer is a water-activated derivative of MDI. As we have discussed, the amino groups on the enzyme react with isocyanate groups on the foam or gel to produce amide.

To prepare the foam-immobilized enzyme, amyloglucosidase was dissolved in a buffer and mixed with the prepolymer. The mixing continued was stopped when the viscosity began to increase. (The molecular weight began to develop. Review our discussion of cream time and tack-free time and the development of an open-cell foam.) The foams were allowed to cure at room temperature. They were then washed with the buffer and squeezed repeatedly. Gels were prepared by the same method. Enzymatic activity was determined by measuring the glucose liberated from various forms of starch.

Apart from the physical nature of the materials, the effects of temperature, pH, and stability were studied. In addition, the immobilization procedures, specifically the buffer to prepolymer ratio, were investigated.

The method described by the authors for immobilizing amyloglucosidase was hypothesized to have several advantages over other methods. The use of the commercially available PU polymers was described as fast, requires no special catalysts, and binds enzymes irreversibly.

The foam immobilization method was thought to have many of the potential advantages:

1) Covalent linkages result in permanent enzyme immobilization.
2) No diffusion limitations were observed in this system.
3) Less product inhibition is seen for the immobilized versus free systems.
4) Enzyme storage stability is enhanced by immobilization.
5) Glucose production at very high temperatures was demonstrated.
6) A wide pH range was conserved.

Overall, PUF-immobilized amyloglucosidase offered several advantages over the free form. The immobilized enzyme is more resilient to long-term storage and the effects of denaturing agents. This directly reduced production costs. The ease of separation of the immobilized enzyme from the product eliminates processing step.

The foams themselves possess several advantageous physical properties. They are flexible and nonreactive once polymerized and offer homogeneous distribution of the immobilized enzyme. They can also be made into a variety of sizes and shapes.

Of particular note were the author's notes on flow-through. The authors suggested that were "amenable for use in column after being frozen in liquid nitrogen and ground to a powder."

While this can be considered a pioneering study in the use of PU prepolymers for immobilization, it reveals the limitation in the technology. This and dozens of similar studies were forced to increase the surface or increase the mass transport through the immobilized PU by artificial and largely mechanical means. Often that meant "cutting the foam into cube" or, as in this case, grinding the foam.

Novel Reactor System

By now you are aware that covalent immobilization of biomolecules is effective and relatively simple. You are also aware of the limitation of this technique with regard to mass transport. The use of the foam as an adsorption device in the University of Maine fish tank experiments and the numerous immobilization studies we have and will continue to cite show the need to increase surface area by "cutting the foam into pieces." We have discussed the problem of discussing immobilization without concern for the architecture of the supporting material. Again while immobilization is necessary research, it is not sufficient in outcome analysis. This is true regardless if the use is medical or environmental. While we have made the case that something approaching the structure of a reticulated PUF, other technologies could be developed that, in the final analysis, improve mass transport.

An interesting technique in this regard was presented by researchers at the University of Idaho [61]. PUs were immobilized with a galactosidase by the standard technique. After polymerization was complete, foam cylinders of 22 mm diameter were cut into slices of 5 mm thickness. The amount of immobilized enzyme was determined gravimetrically.

The PUF structure was examined microscopically. The physical properties of the PUF, such as density and free volume fraction (or porosity), were determined gravimetrically. While not sufficient for general-purpose flow-through applications, it was sufficient for this technique.

The foam was inserted into a syringe. The syringe was designed such that the plunger could be inserted and withdrawn by use of a motor. The syringe was designed such that when the plunger was withdrawn, liquid is pulled into the syringe. When plunger is inserted, the liquid is expelled. Check valves to ensure that the incoming liquid is expelled into a separate container. The expelled liquid was analyzed spectrophotometrically for the products of the enzyme treatment.

The results were guardedly positive with regard to enzymatic activity and half-life. Included in the conclusions were thoughts on the overall efficiency of the prepolymer technique.

Compared with other immobilization methods, *in situ* copolymerization has the following advantages:

1) Immobilization procedures are simple with mild conditions.
2) Prepolymers do not contain monomers that may denature enzymes.
3) Isocyanate groups in the prepolymer react to covalently bind enzymes.
4) The network structure of PUF matrices can be controlled using prepolymers of different chain lengths or by changing the foaming conditions.

Endothelialization

Vascular diseases, especially atherosclerosis, are responsible for a great number of deaths in the United States. A common procedure is to either bypass the occlusion or replace the artery with autologous tissue. Many patients, however, do not have suitable donor tissue. Synthetic materials have been developed for blood vessel substitutes but are not suitable for small diameter applications. Vascular substitutes of less than 6 mm are problematic due to thrombosis and intimal hyperplasia. PU block copolymers have been widely used for biomedical applications due to their excellent mechanical properties and relatively good biocompatibility. Several surface treatments have been evaluated to improve blood compatibility such as heparin and collagen. These modifications, however, have not solved the problem. The physical properties of PU-based vascular grafts would be appropriate if one could improve, for instance, the endothelialization without increasing thrombogenicity.

Researchers at Rice University, Houston, TX, USA, worked to develop a peptide-modified PU to enhance endothelialization for small diameter vascular graft applications [62].

Enhanced attachment, spreading, proliferation, and migration of endothelial cells on vascular graft materials are essential to obtain successful endothelialization. The laminin-derived peptide YIGSR has been used to modify biomaterials to promote endothelial cell growth.

A solution of MDI in 6 ml anhydrous dimethylformamide (DMF) was prepared in 100 ml three-neck round flask and stirred at room temperature. A solution of PTMO in DMF was added and the mixture was heated to 75°C and held there for 2 h under argon gas. The reactor was cooled to room temperature before GGGYIGSRGGGK peptide (0.11 mmol) in 10 ml anhydrous DMF and PPD (1.1 mmol) in 10 ml anhydrous DMF were added as chain extenders. The polymer mixture was incubated at 45°C for 2 h under argon gas. The polymer solution was cooled to room temperature, precipitated in methanol, and dried under vacuum.

Polymers were dissolved in THF (0.3 wt%). The films were held under vacuum for 48 h to ensure removal of the solvent.

The bulk polymer characterization showed that GGGYIGSRGGGK peptide sequences were successfully incorporated into the polymer. The incorporation of the peptides into the polymer backbone did not significantly affect the tensile strength. However, the elastic modulus was decreased and elongation was increased.

Bovine aortic endothelial cells (BAECs) were seeded and cultured on the PU films to evaluate cytotoxicity of any leachables. Over 95% of endothelial cells remained viable at 24 or 72 h. The number of adherent cells was significantly higher than that of control after 4 and 24 h incubation. In addition, cell areas and the percent of spread cells increased with increasing incubation time for both surfaces.

To evaluate ECM production, cells were incubated in the presence of glycine on the PU films. The ECM production was significantly increased compared with the control.

The effect of the peptide sequence on platelet adhesion was examined using whole blood. Platelet adhesion was dramatically lower than on collagen, the control surface. Additionally, there was no significant difference in the number of adherent platelets between the subject film and the control, indicating that the incorporation of the polypeptide did not enhance thrombogenicity.

In conclusion, the PU synthesized with the YIGSR into the polymer backbone showed promising results in vascular graft endothelialization.

Creatinine

The determination of creatinine concentrations in biological fluids is an increasingly important clinical test. This analyte is used for the evaluation of renal function and muscle damage. Creatinine levels in serum and urine are of particular interest because they are not affected by short-term dietary changes.

Researchers have successfully immobilized the three enzymes used in amperometric creatinine biosensors (creatinine amidohydrolase, creatine amidinohydrolase, and sarcosine oxidase) into PU membranes [63].

Hypol prepolymer 2060G (0.4 g) was added to a buffered solution (3.6 g of 50 mM phosphate buffer, pH 7.5) containing creatinine amidohydrolase. The aqueous polymer emulsion was mixed for 30 s, using a spatula, in a weigh boat until the onset of gelation. Polymerization was very rapid and gelation was usually complete within 1 min.

Immobilization of the enzyme into PU improves the enzymes stability in buffer at 37°C, increasing the half-life from 6 days to greater than 80 days. The three-enzyme polymers had considerable stability in buffer, retaining 50% of initial activity after 11 days in buffer at 37°C.

Conclusion to Immobilization

We wanted to conclude this chapter and to some degree the scaffold chapter with two papers developed at Kyushu University, Japan. Not that these are definitive in any way but that they define, in our opinion, a direction that holds great promise. The papers speak more to the architecture than to immobilization, and they open a door to both. The issue of 3D culturing of hepatic cells and by inference islet cells addresses, in our opinion, a settled issue. Still further the development of functioning hepatic spheroids, again in our opinion, is confirmed. Still to be developed, however, is biocompatibility and especially hemocompatibility. Progress on mass transport issues is indicated.

Review of these two papers serves as a summary of both architecture and chemistry and so we conclude this chapter with these two papers. First, we have mentioned earlier [64] that human liver fragments were obtained from patients who underwent hepatic resection. Hepatocytes were inoculated in PUF (1 × 25 × 25 mm). The morphology of hepatocytes in the PUF/spheroid culture system was observed by phase-contrast microscope. Both primary human and primary porcine hepatocytes spontaneously formed spheroids in the PUF stationary culture after 24–48 h of culture (Figure 4.42).

To investigate the metabolic capacity of human hepatocytes in the PUF/spheroid culture, they were cultured in plasma from two patients with fulminant hepatic failure (FHF). After the 24 h culture, the FHF patients' plasma was collected and the concentration of ammonia was measured. The primary human hepatocytes in PUF/spheroid culture showed a significant decrease in the ammonia compared with a monolayer culture. Other factors including amino acid concentrations and bile were measured and compared with monolayer cultures. These other factors were either the same or showed

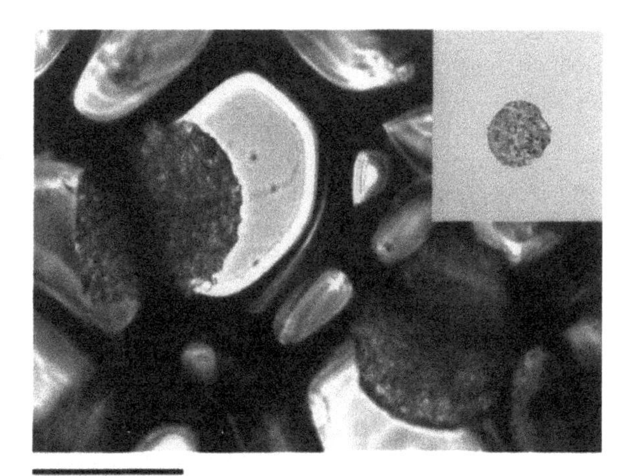

Figure 4.42 Hepatic spheroid cultured in a reticulated polyurethane foam (scale = 100 μm). *Source:* Yamashita *et al.* [9]. Reproduced with kind permission of Dr Yamashita.

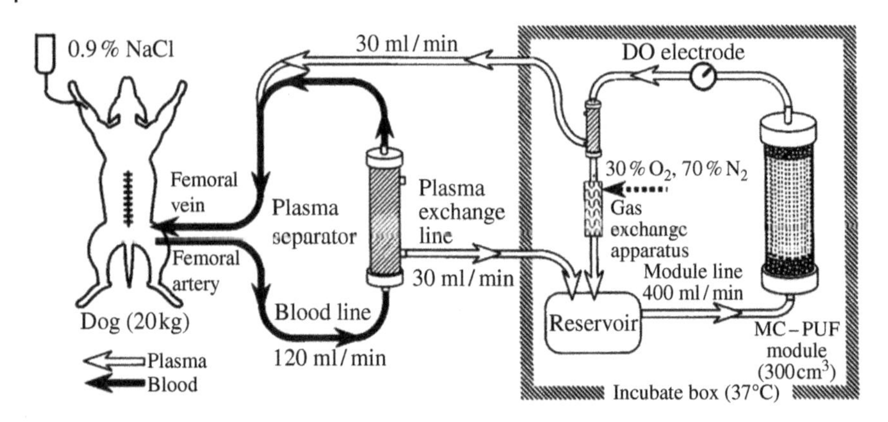

Figure 4.43 Gion *et al.* multicapillary system. *Source:* Gion *et al.* [65]. Reproduced with permission of Elsevier.

improvement. The researchers concluded that the PUF/spheroid culture systems demonstrated the advantageous metabolic function in FHF patient plasma compared with the monolayer culture.

Based in part on this research, Gion *et al.* [65] developed a PUF packed-bed culture system of hepatocyte spheroids as a hybrid artificial liver, which was reported to be effective for recovery from liver failure in a dog acute liver failure model [65].

Hepatocytes were isolated from mongrel dogs of both sexes (weighing 10–20 kg) by the liver perfusion method. Cell viability was determined by the trypan blue exclusion method. A multicapillary PUF packed bed, described by Ijima *et al.*, was used for this study [66]. An extracorporeal circulation system was designed (Figure 4.43).

The blood flow rate was controlled and the total volume of the circulation line. Plasma flowed into the reservoir through a plasma separator at the flow rate of 30 ml/min. Plasma was used as the system was not whole blood compatible. The plasma flow rate of the module line was 400 ml/min, which was required to supply oxygen and nutrients to hepatocyte spheroids. The results of the study show the efficacy of the system. Blood ammonia and serum creatinine levels were significantly lower, and blood pressure and blood glucose levels significantly higher, than those in the control group. For our purposes, however, comments on the architecture were also important. They discussed their research on spheroids of rat and dog hepatocytes immobilized in the pores of PUF as a cell culture substratum. They described that reticulated PUF has a spongelike macroporous structure with each pore about 250 m in diameter made up of smooth thin films and thick skeletons. The hepatocyte spheroids in the pores, about 100 m in diameter, maintain higher liver-specific functions, such as albumin synthesis, ammonia detoxification, urea synthesis, and drug metabolism, compared with 2D cultured hepatocytes for about 3 weeks [67]. Several reports have described features of hepatocyte spheroids and their highly differentiated function [68], in which the metabolic activities was higher than in monolayer-cultured hepatocytes.

In both Matsushita and Gion, in the PUF packed-bed culture system, the spheroids developed spontaneously in the pores of a reticulated PUF. While this was not mentioned, we can only assume that they were hydrophobic. One has to imagine the result of a reticulated foam treated to make it hemocompatible. The critical point, however, is that the hepatic spheroids generally developed spontaneously. Inasmuch as islet spheroids also develop without help, both the liver model and pancreas model would appear to benefit from the use of reticulated PUF.

References

1 *Biofiltration for Air Pollution Control*, Devinny, J.S., Deshusses, M.A., and Webster, T.S., CRC Press, Boca Raton, 1999.
2 *Immobilization of Enzymes and Cells*, ed. Bickerstaff, G.F., Humana Press, Totowa, 1997.
3 Evaluation of an experimental filter medium for water re-use systems, Riley, J., Cole, D., Bauer, R., Department of Biosystems Science and Engineering, Technical Bulletin 177, University of Maine, Orono, ME, September 2000.
4 Protein surface interactions in the presence of polyethylene oxide: I. Simplified theory, Jeon, S.I., Lee, J.H., Andrade, J.D., De Gennes, P.G., *Journal of Colloid and Interface Science* 142 1, 149–158, March 1, 1991.
5 An approach to the targeted attachment of peptides and proteins to solid supports, Hofmann, K., Kiso, Y., Kiso, Y., *Proceedings of the National Academy of Sciences of the United States of America* 73 (10), 1784–1785, 1976.
6 Effects of an avidin-biotin binding system on chondrocyte adhesion and growth on biodegradable polymers, Tsai, W.-B., Wang, M.-C., *Macromolecular Bioscience* 5, 214–221, 2005.
7 Biotin-avidin mediates the binding of adipose-derived stem cells to a porous β-tricalcium phosphate scaffold: mandibular regeneration, Feng, Z., Liu, J., Shen, C., Lu, N., Zhang, Y., Yang, Y., and Qi, F., *Experimental and Therapeutic Medicine* 11, 737–746, 2016.
8 Avidin–biotin binding-based cell seeding and perfusion culture of liver-derived cells in a porous scaffold with a three-dimensional interconnected flow-channel network, Huang, H., Oizumi, S., Kojima, N., Niino, T., and Sakai, Y., *Biomaterials* 28, 3815–3823, 2007.
9 High metabolic function of primary human and porcine hepatocytes in a polyurethane foam/spheroid culture system in plasma from patients with fulminant hepatic failure, Yamashita, Y.-i., Shimada, M., Tsujita, E., Shirabe, K., Ijima, H., Nakazawa, K., *et al.*, *Cell Transplantation* 11, 379–384, 2002.

10 Hybrid artificial liver using hepatocyte organoid culture, Funatsu, K., Ijima, H., Nakazawa, K., Yamashita, Y.-i., Shimada, M., and Sugimachi, K., *Artificial Organs* 25 (3), 194–200.
11 Aqueous polymeric solutions as environmentally benign liquid/liquid extraction media, Huddleston, J.G., Willauer, H.D., Griffin, S.T., and Rogers, R.D., *Industrial & Engineering Chemistry Research* 38, 2523–2539, 1999.
12 The extraction and recovery of polychlorinated biphenyls (PCB) using porous polyurethane foam, Gesser, H.D., Chow, A., Davis, F.C., Uthe, J.F., and Reinke, J., *Analytical Letters* 4 (12), 791–795, 1971.
13 Monitoring of polynuclear aromatic hydrocarbons in water, Saxena, J., Kozuchowski, J., and Basu, D.K., *Environmental Science and Technology* 11, 1977.
14 Selective adsorption of phenols from hydrocarbons solutions using polyurethane foam, Schlicht, R. and Mc Coy, F., US Patent No. 3,617,531.
15 Foamaceous hydrocarbon adsorption medium and method and system for making same, Faudree, T. L., US Patent No. 4,230,566.
16 Preconcentration and separation of some organic water pollutants with polyurethane foam and activated charcoal, El-Shahawi, M.S., *Chromatographia* 36, 318–322, 1993.
17 Molecular basis for some physical properties of alginates in the gel state, Smidsrod, O., *Faraday Discussion of the Chemical Society* 57, 263–274.
18 Function, structure, and stability of enzymes confined in agarose gels, Kunkel, J. and Asuri, P., *PLoS ONE* 9 (1), e86785, 2014.
19 High temperature peroxidase activities of HRP and hemoglobin in the galleries of layered Zr(IV)phosphate, Kumar, C.V. and Chaudhari, A. *Chemical Communications (Cambridge)* (20), 2382–2383, 2002.
20 Current status of clinical islet transplantation, Korsgren, O., Nilsson, B., Berne, C., Felldin, M., Foss, A, Kallen, R., *et al. Transplantation* 79, 1289–1293, 2005.
21 A selective decrease in the beta cell mass of human islets transplanted into diabetic nude mice, Davalli, A.M., Ogawa, Y., Ricordi, C., Scharp, D.W., Bonner-Weir, S., and Weir, G.C.. *Transplantation* 59, 817–820, 1995.
22 Islet encapsulation: strategies to enhance islet cell functions, Beck, J., Angus, R., Madsen, B., Britt, D., Vernon, B., and Nguyen, K.T., *Tissue Engineering* 13, 589–599, 2007.
23 In situ formation and collagen-alginate composite encapsulation of pancreatic islet spheroids, Lee, B.R., Hwang, J.W., Choi, Y.Y., Wong, S.F., Hwang, Y.H., Lee, D.Y., and Lee, S.-H., *Biomaterials* 33, 837–884, 2012.
24 High albumin production by multicellular spheroids of adult rat hepatocytes formed in the pores of polyurethane foam, Matsushita, T., Ijima, H., Koide, N., and Funatsu, K., *Applied Microbiology Biotechnology* 36, 324, 1991.

25 Hydrogels in pharmaceutical formulations, Peppas, N.A., Bures, P., Leobandung, W, and Ichikawa, H., *European Journal of Pharmaceutics and Biopharmaceutics* 50, 27–46, 2000.
26 Pluronic F127 as a cell encapsulation material: utilization of membrane-stabilizing agents, Khattak, S.F., Bhatia, S.R., and Roberts, S.C., *Tissue Engineering* 11, 974–983, 2005.
27 In-situ crosslinkable thermo-responsive hydrogels for drug delivery, Swennen, I., Vermeersch, V., Hornof, M., Adriaens, E., Remon, J.-P., Urtti, A., *et al.*, *Journal of Controlled Release* 116, e21–e24, 2006.
28 In situ formation and collagen-alginate composite encapsulation of pancreatic islet spheroids, Lee, B.R., Hwang, J.W., Choi, Y.Y., Wong, S.F., Hwang, Y.H., Lee, D.Y., and Lee, S.H., *Biomaterials* 33 (3), 837–845, January 2012.
29 A membrane-mimetic barrier for islet encapsulation, Cui, W., Barr, G., Faucher, K.M., Sun, X.L., Safley, S.A., Weber, C.J., and Chaikof, E.L., *Transplant Proceedings* 36 (4), 1206–1208, May 2004.
30 The impact of hyperglycemia and the presence of encapsulated islets on oxygenation within a bioartificial pancreas in the presence of mesenchymal stem cells in a diabetic Wistar rat model, Vériter, S., Aouassar, N., Adnet, P.-Y., Paridaens, M.-S., Stuckman, C., Jordan, B., *et al.*, *Biomaterials* 32 (26), 5945–5956, September 2011.
31 The effects of PEG hydrogel crosslinking density on protein diffusion and encapsulated islet survival and function, Weber, L.M., Lopez, C.G., and Anseth, K.S., *Journal of Biomedical Materials Research* 90 (3), 720–729, 2009.
32 3D co-culturing model of primary pancreatic islets and hepatocytes in hybrid spheroid to overcome pancreatic cell shortage, Jun, Y., Kang, A.R., Lee, J.S., Jeong, G.S., Ju, J., Lee, D.Y., and Lee, A.H., *Biomaterials* 34, 3784–3794, 2013.
33 Impact of polymer hydrophilicity on biocompatibility, Ayala, H.Y., Sullivan, C., Wong, J., Wong, J., David, L., Chen, M., *et al.*, *Journal of Biomedical Materials Research*, 90, 133–141, 2009.
34 Developments in three-dimensional cell culture technology aimed at improving the accuracy of *in vitro* analyses, Maltman, D.J. and Przyborski, S.A., *Biochemical Society Transactions* 38 (4), 1072–1075, 2010.
35 Biomedical applications of polymer-composite materials: a review, Ramakrishna, S., Mayer, J., Wintermantel, E., and Leong, K.W., *Composites Science and Technology* 61 (9), 1189–1224, 2001.
36 Paterson, M. and Kennedy, J, F., in *Methods in Biotechnology, Vol. 1 Immobilization of Enzymes and Cells*, Ed. Bickerstaff, G.F., pp. 153–165, Humana Press, Inc, Totowa, 1997.
37 Swaisgood, H.E. and Passos, F., in *Methods in Biotechnology, Vol. 1 Immobilization of Enzymes and Cells*, Ed. Bickerstaff, G.F., pp. 237–242, Humana Press, Inc, Totowa, 1997.

38 Surface modification of horseradish peroxidase with (poly) ethylene glycol, Laliberte, M., Gayet, J.C., and Fortier, G., *Biotechnology and Applied Biochemistry* 20, 397–413, 1994.
39 Immobilization of collagen onto polymer surfaces having hydroxyl groups, Tabata, Y., Lonikar, S.V., Horii, F., and Ikada, Y., *Biomaterials* 7 (3), 234–238, May 1986.
40 Immobilization of some biomolecules onto radiation-grafted polyethylene beads for possible use in immunoassay applications, Lofty, S. and Moustafa, K., *Journal of Applied Polymer Science* 123, 3725–3733, 2012.
41 Plasma methods for the generation of chemically reactive surfaces for biomolecule immobilization and cell colonization: a review, Siow, S.K., Britcher, L., Kumar, S., and Griesser, H.J., *Plasma Processes and Polymers* 3, 392–418, 2006.
42 Paraffin spheres as porogen to fabricate poly(L-lacticacid) scaffolds with improved cytocompatibility for cartilage tissue engineering, Ma, Z.W., Gao, C.Y., Gong, Y.H., and Shen, J.C., *Journal of Biomedical Materials Research* 67B (1), 610–617, 2003.
43 Glutaraldehyde: behavior in aqueous solution, reaction with proteins, and application to enzyme crosslinking, Migneault, I., Dartiguenave, C., Bertrand, M.J., and Waldron, K.C., *BioTechniques* 37, 790–802, November 2004.
44 The interaction of aldehydes with collagen, Bowes, J.H. and Cater, C.W., *Biochimica et Biophysica Acta* 168, 341–352, 1968.
45 Technical Product Function Sheet, FXI Corp., FS-998-F-5M.
46 Activation of polyvinyl chloride sheet surface for covalent immobilization of oxalate oxidase and its evaluation as inert support in urinary oxalate determination, Pundir, C.S., Chauhan, N.S., and Bhambi, M., *Analytical Biochemistry* 374 (2), 272–277, March 15, 2008.
47 Preparation and use of enzymes bound to urethane, L.L. Wood, US Patent 4,312,946, 1982.
48 4-Oxalocrotonate tautomerase, an enzyme composed of 62 amino acid residues per monomer, Chen, L.H., Kenyon, G.L., Curtin, F., Harayama, S., Bembenek, M.E., Hajipour, G., and Whitman, C.P.. *The Journal of Biological Chemistry* 267 (25), 17716–17721, September 1992.
49 The animal fatty acid synthase: one gene, one polypeptide, seven enzymes, Smith, S., *FASEB Journal* 8 (15), 1248–1259, December 1994.
50 The Catalytic Site Atlas, The European Bioinformatics Institute. http://iubmb.onlinelibrary.wiley.com/hub/journal/10.1002/(ISSN)1470-8744/, Retrieved April 4, 2007.
51 Biochemistry of protein-isocyanate interactions: a comparison of the effects of aryl vs. alkyl isocyanates, Brown, W.E., Green, A.H., Cedel, T.E., and Cairns, J., *Environmental Health Perspectives* 72, 5–11, June 1987.
52 Hydrolyse der insektiziden Phosphorsåureester, Mülhmann, R. and Schrader, G., *Zeitschrift für Naturforschung* 12b, 196, 1957.

53 Esterification activity and operational stability of Candida rugosa lipase immobilized in polyurethane foams in the production of ethyl butyrate, Pires-Cabral, P., da Fonseca, M.M.R., and Ferreira-Dias, S., *Biochemical Engineering Journal* 48, 246–252, 2010.

54 A novel hydrophilic support, CoFoam, for enzyme immobilization, Vasudevan, P.T., Lopez-Cortes, N., Caswell, H., Reyes-Duarte, D., Plou, F.J., Ballesteros, A., *et al.*, *Biotechnology Letters* 26, 473–477, 2004.

55 Cytocompatible polyurethanes from fatty acids through covalent immobilization of collagen, González-Paz, R.J., Ferreira, A.M., Mattu, C., Boccafoschi, F., Lligadas, G., Ronda, J.C., *et al.*, *Reactive & Functional Polymers* 73, 690–697, 2013.

56 Electron microscopic observations on tissue integration of collagen-immobilized polyurethane, van Wachem, P.B., Hendriks, M., Blaauw, E.H., Dijk, F., Verhoeven, M.L.P.M., Cahalan, P.T., and van Luyn, M.J.A., *Biomaterials* 23, 1401–1409, 2002.

57 Prosthesis of Foam Polyurethane and Collagen and uses thereof, Trudell, L.A. and Whittemore, A.D., US Patent 5,207,705, 1993.

58 Adhesion and growth of human umbilical vein endothelial cells on collagen-treated PU/PEGDA IPNs, Yoon, S.S., Kim, J.H., Yoon, J.J., Kim, Y.J., Park, T.G., and Kim, S.C. *Journal of Biomaterials Science, Polymer Edition* 17 (7), 765–780, 2006.

59 Synthesis and properties of poly(ethylene glycol) macromer/β-chitosan hydrogels, Sawney, A.S., Pathak, C.P., and Hubbel, J.A., *Macromolecules* 26, 581, 1993.

60 Immobilization of amyloglucosidase using two forms of polyurethane polymer. Storey, K.B., Duncan, J.A., and Chakrabarti, A.C., *Applied Biochemistry and Biotechnology* 23 (3), 221–236, March 1990.

61 Characterization of immobilized enzymes in polyurethane foams in a dynamic bed reactor, Hu, Z.-C., Korus, R.A., and Stormo, K.E., *Applied Microbiology and Biotechnology* 39, 289–295, 1993.

62 Development of a YIGSR-peptide-modified polyurethaneurea to enhance endothelialization, Jun, H.-W. and West, J., *Journal of Biomaterials Science, Polymer Edition* 15 (1), 73–94, 2004.

63 A stable three enzyme creatinine biosensor. 2. Analysis of the impact of silver ions on creatine amidinohydrolase, Berberich, J.A., Yang, L.W., Bahar, I., and Russell, A.J. *Acta Biomaterialia* 1, 183, 2005.

64 The role of adsorbed fibrinogen in platelet adhesion topolyurethane surfaces, Wu, Y., Simonosky, F.I., Ratner, B.D., Horbett, T.A., *J Biomed Mater Res A* 74(4), 722–738, September 15, 2005.

65 Evaluation of a hybrid artificial liver using a polyurethane foam packed-bed culture system in dogs, Gion, T., Shimada, M., Shirabe, K., Nakazawa, K., Ijima, H., Taku Matsushita, T., *et al.*, *Journal of Surgical Research* 82, 131–136, 1999.

66 Development of a hybrid artificial liver using a multicapillary PUF/spheroid packed bed, Ijima, H., Matsushita, T., and Funatsu, K., *Japanese Journal of Artificial Organs* 23, 463, 1994.

67 Hepatocyte spheroids in polyurethane foams: functional analysis and application for a hybrid artificial liver, Ijima, H., Matsushita, T., Nakazawa, K., Fujii, Y., and Funatsu, K. *Tissue Engineering* 4, 213, 1998.

68 Extended liver-specific functions of porcine hepatocyte spheroid entrapped in collagen gel, Lazar, A., Mann, H.J., Remmel, R.P., Shatford, R.A., Cerra, F.B., and Hu, W.S., *In Vitro Cellular & Developmental Biology—Animal* 31, 340, 1995.

5

Controlled Release from a Hydrogel Scaffold

Introduction

This may seem like a divergence from the main theme, but it is appropriate to consider a scaffold, as we have been describing, not only as a support or surface for immobilization but also as a scaffold for the delivery of active compounds, including proteins. We have discussed the immobilization of proteins by covalent bonding. This permanently attaches the protein, but with the techniques of adsorption and even encapsulation, under the proper conditions, they can diffuse from the scaffold into its environment.

An example illustrates. We have discussed the adhesion of osteoblasts to the chitosan/PU scaffolds [1]. In the following study, chitosan was used as a delivery system for insulin [2]. It was chosen for the study because of its biocompatibility and biodegradability. The polysaccharide was dissolved in acetic acid and insulin was added. Capsules were formed by dispersing the solution in paraffin oil, followed by a cross-linking agent. The report goes on to demonstrate the release of insulin to a biological fluid and blood.

The point of mentioning it here is as a segue to the discussion of scaffold, not as an immobilization system for biomolecules and cells, but for an interesting and potentially important role in the biological process. We have discussed various degrees of immobilization, from adsorption to encapsulation (the example of chitosan previously) in which the potential exists and that the active can diffuse from the scaffold for some purpose, beneficial or toxic.

We would like to explore adding another attribute to the general topic. We described at length the construction and purpose of a scaffold. In that regard, the architecture and strength of materials were the goal. We have at many times expressed our prejudice toward flow-through systems and our liver

Polyurethane Immobilization of Cells and Biomolecules: Medical and Environmental Applications, First Edition. T. Thomson.

model. Still further, we have discussed reticulated foam and composites thereof. In the latter, we have expanded the applicability by suggesting coating the otherwise hydrophobic foam surface with biocompatible components.

In discussing immobilization, particularly in the case of covalent attachment, our goal was to permanently change the surface energies. In the case of enzymes, specifically, the orientation on the surface at appropriate binding sites was appropriate for extending the half-life.

By contrast, however, we suggest that there are reasons why one might want to consider the scaffold that we have carefully developed as a delivery system for beneficial molecules. We will explore this concept, but it is important for the theme of this book that we discuss it within the context of an appropriate scaffold. For this reason we will use a model for our exploration. There are many methods of controlled release, and many are not appropriate for scaffold applications. By way of example, biodegradable polymers are useful as release devices, but combining this with its function of a scaffold would appear to overcomplicate the overall function.

In this exploration, we will use the concept of a biodurable scaffold with a biocompatible surface. Again while there are many clever systems for the delivery of actives, we will limit our discussion to a useful technology for delivery, diffusion. Consider the following Figure 5.1. It is a construction composed of a scaffold, a reservoir layer, and a biocompatible layer. The reservoir layer and perhaps modifications of the biocompatible layer are the topic of this discussion.

Controlled release systems for active molecules offer several advantages. While not necessary, the techniques described here offer an option. The use of conventional controlled release is to extend the delivery of an active for an extended period of time in the therapeutically significant range. Let us say, for

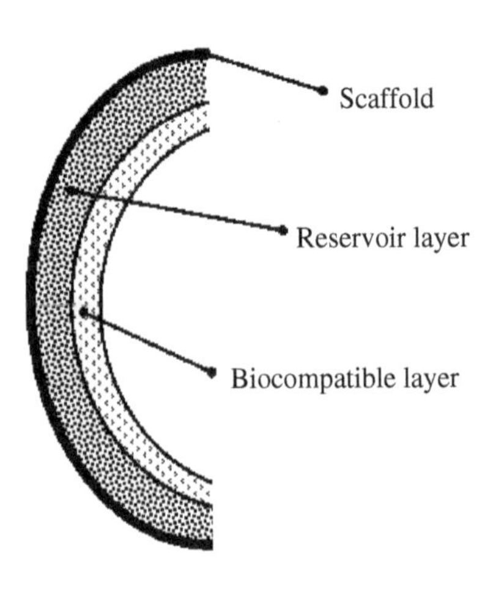

Figure 5.1 A segment of a scaffold pore showing a reservoir layer and the biocompatible surface.

instance, that we have developed a scaffold like the chitosan/insulin described earlier. By injection, we would expect an instant increase in concentration followed by an equally rapid decrease. By imbedding it in the chitosan scaffold, however, the release of insulin can be extended. Figure 5.1 illustrates this in a qualitative manner.

There are a number of methods used to control the delivery of actives. The following list shows the range of techniques. As this book is about immobilization and not delivery, we won't cover these in detail with the exception of one, diffusion:

- Membrane-controlled reservoir systems
- Diffusion-controlled monolithic systems
- Biodegradable systems
- Osmotic systems
- Mechanical systems

In Figure 5.1 is a discussion of a system in which a reservoir of active ingredient is released into an environment through a rate-controlling polymer (the biocompatible surface). Each of these components has specific requirements and constraints, and while these properties overlap, they have important differentiating characteristics. As we said, it is our intension to explore the construction shown in Figure 5.1. This would be referred to as a diffusion device with a rate-controlling membrane. The membrane is the biocompatible surface, which may include immobilized biomolecules or cells as discussed in the last chapter. In this chapter we add another option. By this technology, one can release a useful component(s) that would be useful to the process, for example, nutrients or growth factors.

There are a wide variety of applications for controlled delivery. Applications range from fatty acids as anti-inflammatory agents for release in the intestines to flavor ingredients in food products. Lipophilic drugs can be encased in a hydrogel released in the stomach, the small intestines, or the colon, depending on the intended effect.

In all cases the design of the delivery system must include the predictable release pattern, as we illustrated in Figure 5.2, while the primary goal is to extend release. At least two dependent variables need to be considered. The dose by injection must be therapeutically significant without exceeding the toxic limit. The type of the controlled release attempts to extend the time within those boundaries.

Before we get to the details and examples, there is a host of materials used for controlled delivery and many methods of preparations. We will go through these and show that many of these chemistries were also described in the scaffold chapter and with regard to immobilization. We used the aforementioned chitosan examples as an illustration. Before we get to that, however, we need to discuss release patterns in more detail.

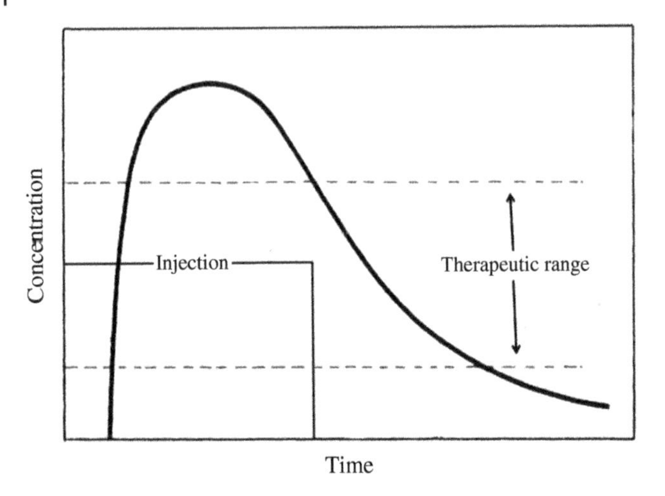

Figure 5.2 Comparison of injection and a controlled release device for insulin.

Release Rates

The release pattern from a device can vary over a broad range. The pattern is typically described by the shape of the release curve. The simplest of these is the so-called "zero-order" release. In a sense it is the goal of almost all controlled release systems. For physical reasons, however, it is rarely achieved except in mechanical systems. The second pattern is probably the natural release profile. It is called a "first order release," which is the result of a release rate based on the difference between the concentration in the device and the concentration in the environment. A third pattern is referred to as a "square root release." The best example is when an active ingredient is encased in a degradable or erodible matrix.

These patterns are illustrated in Figure 5.3.

Examples of Hydrogels Used for Controlled Release

The following list is of hydrogel materials that are regarded as safe in general controlled release applications. Our application, however, is far from general. Nevertheless this gives some guidance as to toxicity. You will notice that many have been discussed as scaffolds. A device that is developed in connection with our investigation will be guided by additional considerations. Nevertheless, in the context of controlled release, it would appear to be worthwhile.

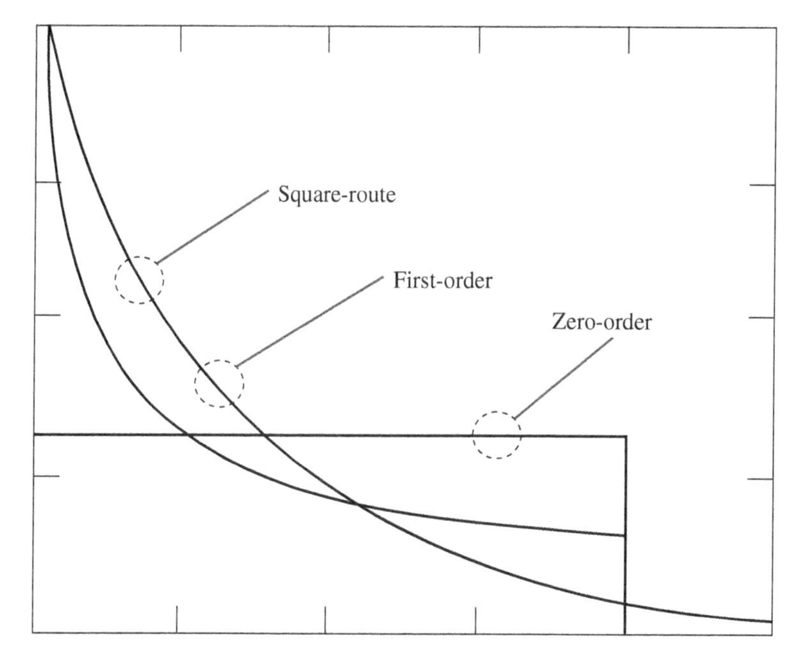

Figure 5.3 Rates of controlled release.

In the following discussion, the following materials are thought of as the reservoir layer. As such they are typically converted into hydrogels by the method described.

Polysaccharides

These are typically derived from plants by microbial processes and vary in many of their important properties, including molecular weight and branching. They can be converted into hydrogels by a number of processes.

Pectin

Pectin is also known as pectic polysaccharides. Several distinct polysaccharides have been identified within its structure including galacturonic acid. Pectin has a molecular weight of typically 60–130 000 g/mol, varying with origin and extraction conditions. It is converted to a hydrogel by two methods; we are most familiar with the addition of sugar. Gelation is also achieved by adding divalent salts, calcium, for instance. Pectin is not digested by gastric or small intestine enzymes, but is easily degraded by organisms in the colon.

Alginates

Alginates are a family of naturally occurring polysaccharides. It is extracted from brown seaweed and used by the pharmaceutical industry for specific gelling, thickening, and stabilizing applications.

Alginate hydrogels are formed by the cross-linking action of divalent cations (Ca^{2+}) as we discussed in an earlier chapter. Alternatively cross-linking is achieved by the addition of carboxylic acids. The latter can be used to affect novel release characteristics.

Carrageenan

Carrageenans are also a linear polysaccharide. Hydrogels are made by adding a divalent salt, but then the system must be cooled. By this process the material undergoes a coiling that results in the hydrogel. Three forms of the polymer exist. Some gel with calcium, while others react in the presence of potassium ions.

Agar

Agar is derived from the polysaccharide agarose, which forms the supporting structure in the cell walls of certain species of algae and is released by boiling. Agar is actually the resulting mixture of two components: the linear polysaccharide agarose and agaropectin. It can be converted to a hydrogel by first heating, followed by cooling. A coil helix configuration and hydrogen bond is the apparent cause. Agar hydrogels are rigid and brittle.

Starch

Starch is a water-soluble polysaccharide composed of amylase and amylopectin. No special treatment is needed to produce gels. Certain types of starch are degraded in the mouth and stomach, whereas others can reach the colon. Modified starches have been used for the delivery of lyophiles.

Proteins

Protein-based hydrogels are commonly used as delivery systems. Whether by heat-set, cold-set, or ion-set, they present interesting biocompatible options. After fabrication, the hydrogels can be can be hardened by several chemical methods.

Gelatin

Gelatin is derived from the processing of animal collagen. Commercially, this is most typically obtained from cattle hides and bones and pigskins.

In its most basic form, gelatin is a tasteless beige or pale yellow powder or granules. It is composed of mostly protein, with a small percentage of mineral salt and water, making up the balance. Hydrogels are the result of non-covalent cross-links induced by cooling below the coil helix transition temperature.

Casein

Casein is derived from milk by several processes including enzymatic treatment. Caseins have relatively disordered and flexible structures with high surface activities. They are often used as emulsifiers for fatty substances. Because of their globular disordered conformation, they are stable against thermal variations.

Other Proteins

Among these are albumin and egg proteins. They are typically soluble in water but can form hydrogels in alcohol solutions.

Controlled Release by Diffusion

There is a wealth of literature on the subject, but we need to stick to the model of a reservoir applied to or developed in the process of building a scaffold. The object of this discussion is to create a physical environment in which an active ingredient can reside on till needed. The release of the active ingredient can be by any of several methods and it is not our purpose to fully describe the subject. We have a physical model with which we will work, and as such we will focus on a reservoir layer and a rate-controlling membrane.

In diffusion, a substance is released from a device to the environment. In our case, diffusion will be from a reservoir into a membrane and then into the environment. While each segment is similar in nature, there are important differences. The membrane is a biocompatible surface to which cells can propagate or has been modified with biomolecules. Its primary and secondary purposes are to accommodate organisms or molecules and to serve as a controlled release system.

The purpose of the reservoir layer is to store and deliver a substance that supports the primary goal of the membrane. We will discuss these two systems. Be aware, however, that the subject is more complex than what we will cover here. There are many texts that are complete and certainly more qualified than ours. Among them is a work by Baker in 1987, which we highly recommend [3].

Reservoir Layer

In the reservoir the active ingredient is typically dispersed throughout a rate-determining matrix. For the purpose of this discussion, consider the matrix to be a hydrogel. While this system works for our discussion, it is not universally necessary. There is a commercial shaving product that has a reservoir layer composed of a polyvinyl chloride in which is dispersed into a polyethylene oxide (PEO) powder. Upon wetting the PEO dissolves and acts as a lubricant. In another case an insecticide is incorporated into a PVC dog collar to dispense flea-killing chemicals.

The shape of a release curve is the most important to us. We illustrated the general categories, but it is diffusion that interests us most. If a solute is dissolved in a hydrogel and then placed in water, the shape of the curve will roughly follow that pattern of Figure 5.1. While this is true, we are interested on the quantitative values, which are dependent on the specifics of the hydrogel and the active ingredient.

Before we tackle that, however, let us return to our model. We have described the reservoir layer to a sufficient degree, but now we must juxtapose it with the membrane layer. At time zero we will assume that the membrane layer has no active ingredient. In this description the reservoir and the membrane are identical in every way. When it comes into physical contact with the reservoir layer, migration begins. At the reservoir/membrane surface the active ingredient is absorbed to the same concentration. If the membrane and reservoir are identical in all respects, a concentration front will develop, move across the membrane, and eventually reach the environment. The process effectively dilutes the active ingredient and, as the front reaches the environment (Figure 5.4), its release begins.

If, however, the nature of the reservoir and the membrane are different in any of a number of ways, a more complex pattern develops. The difference is in the nature of diffusion. To explain we apply the most common relationship,

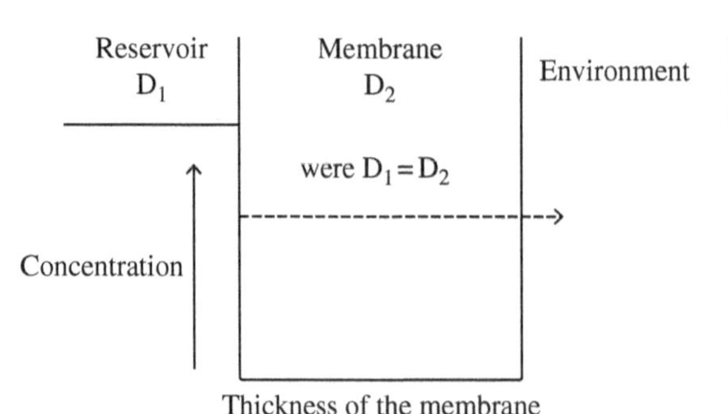

Figure 5.4 Migration of the solute from a reservoir into an identical membrane.

Fick's law of diffusion. It explains that solutes diffuse from regions of high concentration to regions of low concentration and that the rate is proportional to the difference in concentrations:

$$J = D\frac{dC}{dx}$$

where

- J is the diffusion rate (the amount of solute that will flow through a unit area during a unit time interval).
- D is the diffusion coefficient.
- dC/dx is the concentration gradient.

Without going into any more detail than that, it is clear that we need to examine the diffusion coefficient. As we have described we set up a construction in which the reservoir layer containing a solute is in intimate contact with a membrane layer. Again Figure 5.4 describes the result if the two layers are identical. In this context it is only identical in the sense that the diffusion coefficient is exactly the same. This poses the question, however, what if they aren't and more specifically what if diffusion constant is less in the membrane layer than in the reservoir layer. In the context that our membrane layer is designed based on the needs for a biocompatible surface with immobilized biomolecules, we can certainly anticipate that it will be different. Figure 5.5 shows what might be the result. It is therefore appropriate that we address the factors that affect "D."

Einstein expanded his work in collaboration with Stokes on the settling rate of particles in a liquid and applied similar principles to diffusion. What resulted was the Stokes–Einstein equation. It imagines a sphere moving through a continuous fluid; the driving force is to remove the thermodynamic inconsistency of nonuniform concentration gradients.

Our interest, however, is not in fluids, but rather in gels. To apply the relationship, however, we must differentiate the "fluid" with regard to the macroscopic property of a gel and the microscopic viscosity. Hydrogels by definition

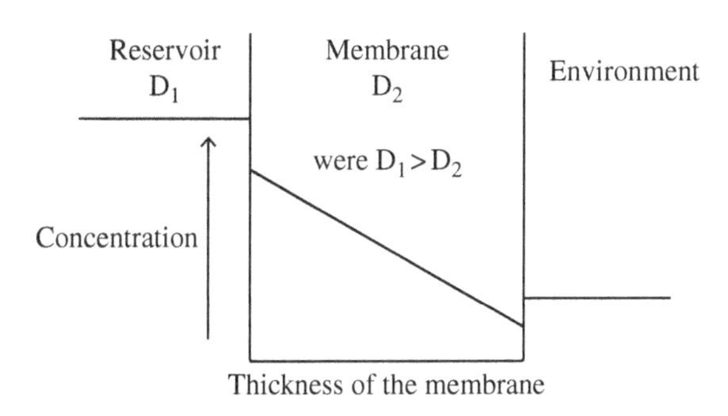

Figure 5.5 Migration of a solute from a reservoir into a membrane with a lower diffusion coefficient.

have infinite viscosity and therefore would not appear to apply. The microscopic viscosity is the intermolecular spaces between the polymer chains. Those spaces are of course water through which the solute can travel. As we develop this argument, we will discuss this phenomenon in relation to the molecular size of the solute and the extent to which the water is affected by the polymer. Thus, according to the Stoke–Einstein relationship water acts as a fluid to solutes passing through them. In fact it is fairly good predictor of diffusion coefficients even in a system of infinite macro-viscosity.

As we mentioned, the size of the spheres (as measured by the MW of the solute, for instance) as one might suggest has a strong effect on the diffusion coefficient (Figure 5.6). We must realize that the solutes with which we are interested are those that have a biomedical significance. In this sense, they are predicted by the application and are not chosen for diffusion reasons. Thus the hydrogel must be designed to allow the diffusion of the solute to be sufficient, not the other way around. Nevertheless, it is important to recognize the molecular size of the solute [4].

It is the gel, therefore, that is the object of our development effort. In this context we have two to be designed. The reservoir layer should have high diffusivity, while the control of the release from the membrane should deserve additional attention. One can imagine that a low polymer concentration of relatively low chain lengths, while still able to develop a gel, might approach the Stokes–Einstein relationship. When one includes the concept of polymer cross-linking, it might move in the opposite direction. A complex three-dimensional structure would intuitively hinder solute mobility and thus decrease the diffusion coefficient. To explore this we will examine work at the University of Colorado and by Prof. Laney Weber *et al.* [5].

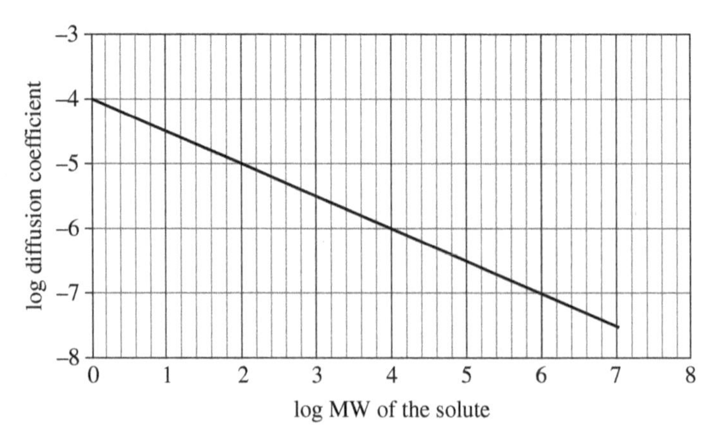

Figure 5.6 Diffusion coefficient as a function of molecular size.

In continuity with the general purpose of this text, they recognized that cell and tissue replacement strategies involve a number of independent technologies. Applying this to islet insulin-producing cells involves both survival and diffusion characteristics, not to mention scaffold architectures. As we have discussed hydrogels and particularly polyethylene glycols (PEGs) are considered an appropriate material for development in these regard. In the present paper, gels are made by the photopolymerization of diacrylate derivatives of PEG. Whether by photopolymerization of diacrylates or by incorporation into polyurethane, the functional part of the molecule is dominated by the glycol backbone. It is worth noting, however, that photopolymerization produces radicals that cause the cross-linking. This makes the result less predictable than the covalent-controlled cross-linking in polyurethanes. The amount of cross-linking, however, is also the subject of this study. Regardless of the mechanism, the intent is to show the effect of cross-linking on the diffusion coefficient so as to add this as a critical design feature of the membrane and reservoir. In the later case we will want minimum cross-linking to achieve sufficient strength, but with the membrane the release rate of a solute needs to be controlled. To repeat a concept discussed earlier, the extracellular matrix releases growth factors and enzymes at a controlled rate, and our synthetic membrane should also have that property. In the present paper, the release of insulin-producing factors must be controlled as well.

The subject paper involves the determination of the diffusion coefficient in a series of PEG gels at what they report to be various cross-link densities. The photopolymerization is described in Lin-Gibson [6]. Molecular weights of the glycols ranged from 2000 to 10 000. Hydrogels were produced from aqueous solution using a photoinitiator and exposed to 365 nM light. Disc-shaped coupons were made at 4 mm thickness. The volumetric swelling was determined and used to measure the cross-linking density (Figure 5.7).

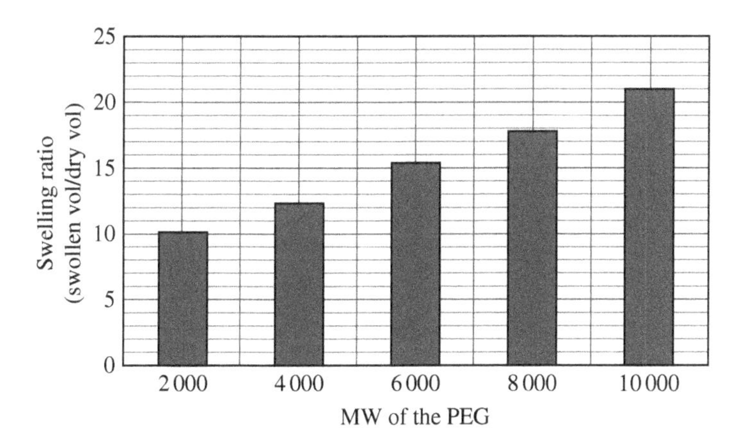

Figure 5.7 Volumetric swelling in serum (Ref. [5]).

The text goes on to estimate the degree of cross-linking using "the theoretical concentration of cross-linkable double bonds in the hydrogel precursor solution." For reasons discussed earlier, polyurethane chemists have a more quantitative way to measure cross-linking density (the amount of triol, for instance) and so while Weber is justified in his interpretation, we think the effect has more to do with the "soft segment" concentration. While "theoretical calculations" have their place, as you will see, it is not necessary. We will propose a more intuitive interpretation that arrives at the same conclusion but in a form that the experimentalist can use to design an experiment. The conclusions are the same, but the interpretation is different. This is not to say that the degree of cross-linking doesn't play an important role. It is, of course, what separates a gel from a high viscosity liquid.

Looking at it more fundamentally, the higher the volumetric swell, the higher the water content of the gel. This would appear to be a more intuitive explanation for the data. While cross-linking probably plays a role, it is hidden within the water effect. Again, in a polyurethane context, one would attempt to limit the swell by increasing the triol, thus allowing for the determination of the diffusion coefficient as a function of cross-linking at a normalized water concentration.

If the same experiment was conducted with the same polyol range in a polyurethane context, an equivalent data set could be developed. In that experiment, several levels of triol at each molecular weight could be used. Diffusion coefficients could then be extracted without a relationship to molecular weight.

As an aside, a hydrogel can be made from a foaming hydrophilic polyurethane, thus allowing the experimenter the ability to do hydrogel experiment similar to the Weber study. If the prepolymer is dissolved in acetone and then water is added, the viscosity of the solution is low enough for the carbon dioxide to escape. The result is a dense gel.

Diffusion Experiments

In Weber, the following proteins were used as solutes: insulin, myoglobin, trypsin inhibitor, carbonic anhydrase, ovalbumin, and bovine serum albumin. The measurement of diffusion was conducted by incubating each of the samples in serum. At intervals, the coupons were transferred to fresh solutions. The amount of protein was measured in each solution, and this was used to calculate the diffusion coefficients (Figure 5.8).

As you can see, the diffusion coefficient increases with the molecular weight of the original polyol. A similar graft could have been made of the diffusion versus the swelling ratio without regard to the cross-link density (Figure 5.9).

The hydrodynamic radius is related to the molecular weight and other factors. One would expect that as the radius increases, the diffusivity would decrease. This is confirmed in Figure 5.10.

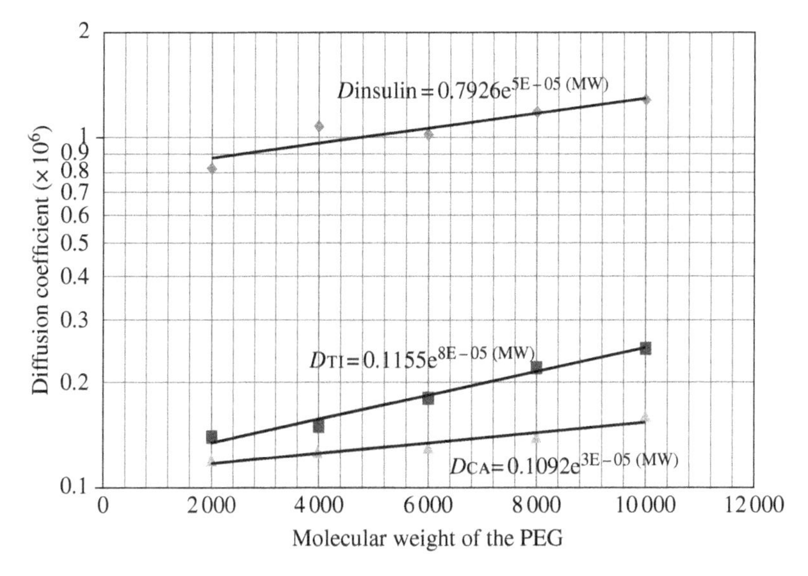

Figure 5.8 Diffusion coefficient as a function of the molecular weight of the PEG. *Source:* Data extracted from Weber et al. [5]. (*See insert for color representation of the figure.*)

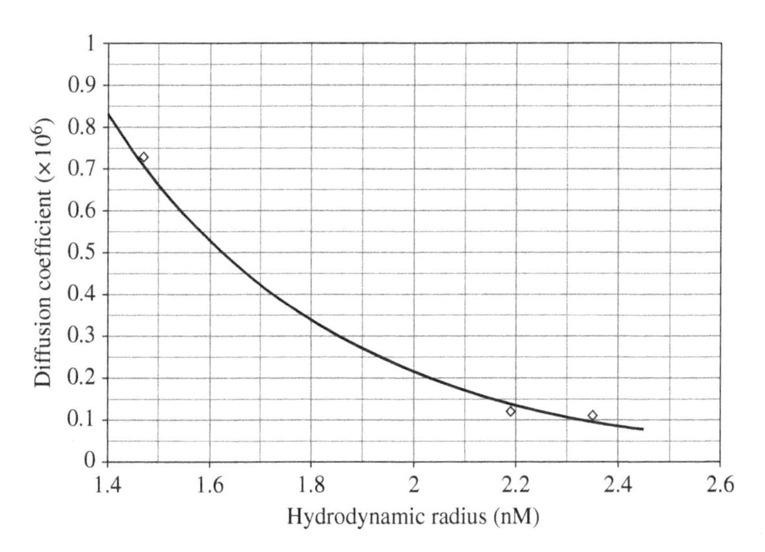

Figure 5.9 Relationship of the diffusion coefficient and the hydrodynamic radius. *Source:* Data extracted from Weber et al. [5]).

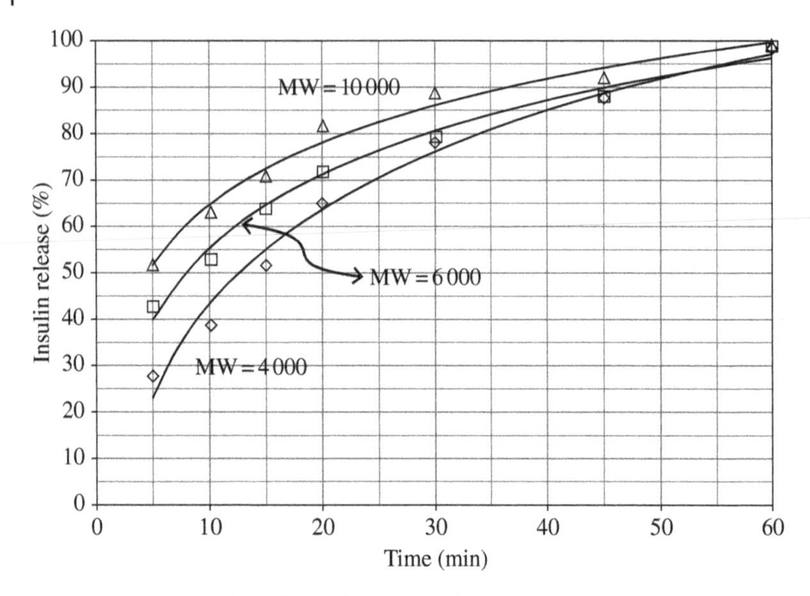

Figure 5.10 Insulin release from islets encapsulated in PEG gels. *Source:* Adapted from Weber *et al.* [5].

In conclusion while our interpretation is different, the information in Weber allows us to confirm several important points concerning

- The effect of water content on diffusion
- The relationship of soft segments on swelling
- The effect of molecular size on the diffusion of proteins

Again this qualitative analysis is done without regard to the level of cross-linking.

The paper goes on to show the diffusion of insulin from islet cell encapsulated in a PEG hydrogel. This returns us to our general theme of immobilization and scaffolding.

Islet Encapsulation

Islets from adult mice were cultured at 37°C under humid conditions with 5% CO_2. Approximately 20 islets were suspended in 30 μl of hydrogel precursor solution, followed by photopolymerization as described previously, entrapping the islets within the gel network. Swollen, islet-containing hydrogels were approximately 4 mm in diameter and 1 mm in thickness. Cell viability was

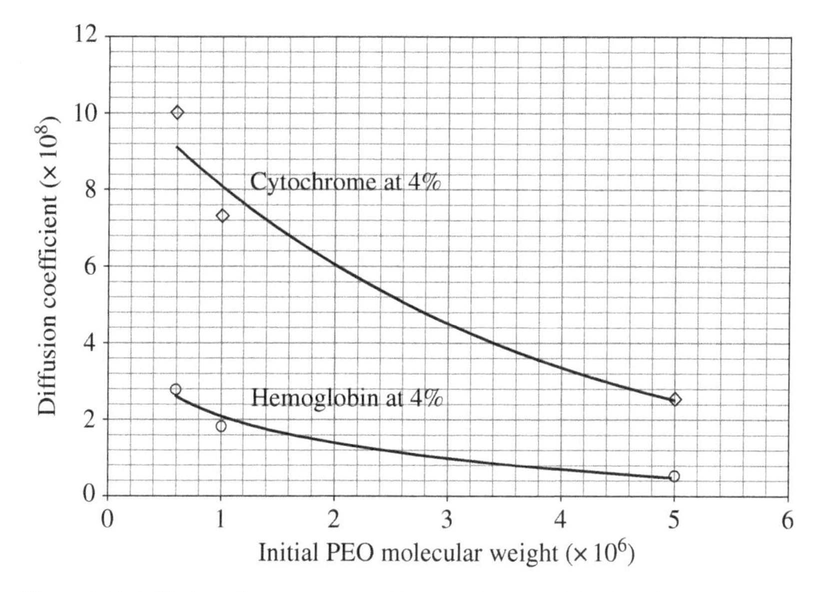

Figure 5.11 Diffusion of cytochrome and hemoglobin in PEO hydrogels.

determined by staining. Insulin secretion was evaluated by exposure of the encapsulated islets to glucose. Samples were placed in high glucose solutions (250 μl each) at 2, 5, 10, 15, 20, 30, 45, and 60 min. The insulin content of each solution was measured, and cumulative insulin secretion was plotted as a percentage of the total amount secreted (Figure 5.11).

Concluding our comments on Weber, this discussion adds the aspects of controlled release through diffusion to the immobilization by encapsulation. Again, higher molecular weight and the associated swelling of the gel increased diffusion. As an aside, one might imagine building a reservoir similar to Weber (but with polyurethane) on a flow-through scaffold would appear to be a fruitful area for research.

The conclusions of the Weber paper support the concept that a PEG hydrogel can be a useful reservoir for proteins and/or a membrane to control the release (see Figure 5.1). Earlier chapters describe it as a hydrophilic if not biocompatible surface for colonization and immobilizations.

There is a variation of PEG that also might be considered. The following paper describes the same polymer but a much higher molecular weight. PEG, as we have described them, are commercially available in molecular weights up to about 10 000. There is an identical molecule, made by another process that reaches millions. As such it is referred to as polyethylene oxide and is sold by Dow Chemical among others. Molecular weights of up to 5 million are available. It is soluble in water. Our involvement was as a wound dressing; the purpose of which was to hydrate the wound bed and as an additional benefit had a cooling effect. In order to make the dressing, a PEO solution was exposed to an electron beam, thereby cross-linking the polymer, producing a hydrogel.

Merrill *et al.* used that technology to study the diffusion of proteins and found what they felt was an unexpected relationship among the concentration of the PEO and diffusion. In the context of Weber, however, perhaps it should not have been a surprise. In any case, the data is valuable to our purpose.

Three molecular weight ranges were used (600 000, 1 million, and 5 million) and solutions of 2, 4, and 6% were made from each. The solutions were exposed to three million electron volts from a van de Graaff generator and then cut into coupons of a convenient size. Swelling measurements were made and correlated to the average molecular weight between cross-links. We assume the calculations were similar to the methods used by Weber. The analysis of this, while in our opinion not necessary for our purpose, is interesting and should be examined.

A small hole was cut into the center of each coupon to which is placed with a cytochrome *c* or a hemoglobin solution. The diffusion of the proteins was measured colorimetrically. The coefficients were calculated and are restated here in a graphic form in Figure 5.11.

The effect is opposite to the Weber study. Molecular weight appears to have a negative relationship to diffusion. Swelling data is not available, but I think there is an intuitive answer. Remembering that these molecular weights are orders of magnitude higher than those in Weber, and that these are very hydrophilic polymers; our interpretation is based on the hydrodynamic environment. All hydrophilic polymers develop a hydration shell. If the molecular weight is not high enough, it does not violate the model of a solute passing through a low microscopic viscosity. At the same concentration at higher molecular weight, however, the structure building property of PEO in this context is sufficiently strong to inhibit diffusion. Again this is from an experimentalist view, not that of a physical chemist with math skills.

One more point, these two experiments were not able to differentiate the effect of cross-linking, in our opinion. Both would have benefited by a simple measurement of the gel strength and constant water.

At this point, I have to make a personal comment. My work has always been as an experimentalist (even if I didn't do the work). As such, data must be in a form that is translatable into quantitative measurements. There is a tendency in my field, physical chemistry, to overcomplicate technology with mathematics. As my professor would tell me, you can graph a giraffe if you add enough functions to an equation. This is a practical guide to bench level research and while developing models is fun (I have fallen in this trap), it doesn't help when you have some beakers or tanks in front of you. Both of the last two papers represent excellent laboratory skills and data collection, but to overcomplicate the issues with math, in my opinion, does not add to the science. Indirectly they recognize that in their concluding figures. Finally, mathematics is equivalent to medical scientists speaking of the "bolus of biomass."

In these two experiments, as designers and experiments of a controlled release hydrogels to support our scaffold project, we know that the amount of swelling increases diffusion (from Weber), and as the length of the polymer chain increases, diffusion decreases.

Other Controlled Release Examples

Again focusing on the release of biomolecules to the environment without regard to other factors, this discussion is an expansion of the scaffold concept. To explain, consider the release of lipophilic bioactive agents into the gastrointestinal tract (GIT). The agents need to be encapsulated within delivery systems to overcome problems associated with delivery. Hydrogel particles are useful for encapsulation, protection, and controlled release of lipophilics. The application of hydrogel particles for release in the gastrointestinal tract includes the mouth, stomach, small intestine, and colon. Hydrogel particles may be designed to protect bioactive components from metabolism or digestion during passage through specific regions of the GIT. For example, the digestion of triglycerides within the stomach and small intestine can be retarded by encapsulating them [7]. As we discussed earlier, hydrogel particles may be designed to control the diffusion of encapsulated bioactives. They may be designed for sustained release depending on their composition, size, structure, and environmental responsiveness [8].

Targeted Delivery

This may seem disconnected from our general discussion, but the following provides another technique to effect control over the release of bioactives. Zhang *et al.* produced an excellent review of the targeted release of bioactives from a hydrogel scaffold within the GIT. After a generalized discussion of hydrogel scaffolds, the paper focused on specific sections of the tract.

Hydrogel particles can be designed to retain, protect, and deliver bioactive components to specific regions within the GIT. They can be designed as a protection from conditions in the stomach and then release them in the gut. The physiological conditions in the several regions of GIT are required to deliver active ingredients.

Stomach

It may be appropriate to release an active ingredient in the stomach to enhance its bioactivity. The environment, however, is characterized by acidic gastric fluids, high mineral content, and digestive enzyme. Still the stomach is the direct route into the blood stream. Calcium alginate beads have been shown to remain intact under stomach conditions and are therefore capable of protecting bioactive components in the gastric environment. Protein-based hydrogel particles have been shown to degrade under simulated gastric conditions and may therefore be suitable for releasing bioactive components in the stomach [9].

Small Intestines

Beginning in the duodenum, hydrogel particles will see a different environment. Alkaline intestinal fluids will act on the hydrogel. Pancreatic enzymes in the small intestine will continue to metabolize residues. This is particularly important for oil lipophilic molecules because of their low solubility in gastrointestinal fluids. These components need to be released from the hydrogel particles within the small intestine and then solubilized within micelles formed by bile salts and phospholipids. The micelles transport the encapsulated bioactive components to be absorbed in the walls of the gut.

Li *et al.* used chitosan/calcium alginate hydrogel particles to encapsulate protein-coated lipid droplets [8]. A simulated model showed that the hydrogels could control the lipid digestion rate. Mun *et al.* showed that β-carotene bioaccessibility was increased after being encapsulated in a starch-based hydrogels [10].

Colon

Colonic delivery imposes additional requirements. Bioactive components within hydrogel particles must pass through the other regions of the GIT without being released or absorbed. Dietary fibers are digested in the lower GIT due to enzymes secreted by colonic bacteria. This makes them suitable for colonic delivery systems. Several dietary fibers had been investigated for colon-specific hydrogel particles such as pectin, alginate, dextran, guar gum, and chitosan.

Fat malabsorption is common in individuals with cystic fibrosis and pancreatic insufficiency. This places them at risk for caloric, essential fatty acid, and choline deficiency, which may in turn lead to growth failure and a poorer clinical course.

A scaffold referred to as an "organized lipid matrix" (OLM) was used in a feeding study. The OLM was composed of phosphatidylcholine and phosphatidylglycerol cholesterol combined in a proprietary process. The material led to the development of Lym-x-Sorb® lipid matrix (BioMolecular Products, Haverhill, MA).

The OLM led to better clinical outcomes in terms of energy intake from the diet, weight-for-age *Z*-score, essential fatty acid status, vitamin E, and retinol-binding protein. These results suggest that OLM is a readily absorbable source of fat and energy in CF and is an effective nutritional supplement [11].

Summary and Conclusions

Hydrogel particles can be used to encapsulate a wide range of bioactive components. They are also in the food industry and are used to protect bioactive components from digestion in specific regions of the GIT. While this work does not relate directly to the topic of scaffolding and immobilization, we feel, as a method for controlled release in vivo, it is a tool we need to understand.

References

1 In vitro evaluation of polyurethane-chitosan scaffolds for tissue engineering, Imelda Olivas-Armendariz, I., Perla García-Casillas, P., Estradal, A., Martínez-Villafañe, A., De la Rosa, A., and Martínez-Pérez, C.A., *Journal of Biomaterials and Nanobiotechnology* 3, 440–445, 2012.

2 Chitosan microcapsules as controlled release systems for insulin, Aiedehm, K., Gianasi, E., Orienti, I., and Zecchi, V.M., *Journal of Microencapsulation* 14 (5), 567–576, 1997.

3 *Controlled Release of Biologically Active Agents*, Baker, R.W., John Wiley & Sons, Inc, New York, 1987.

4 Controlled release: mechanisms and rates, Baker, R.W. and Lonsdale, H.K., in *Controlled Release of Biologically Active Agents*, Ed. Tanquary, A.C. and Lacey, R.E., pp. 15–72, Plenum, New York, 1974.

5 The effects of PEG hydrogel crosslinking density on protein diffusion and encapsulated islet survival and function, Weber, L.M., Lopez, C.G., and Anseth, K.S., *Journal of Biomedical Materials Research* 90 (3), 720–729, 2009.

6 Synthesis and characterization of PEG dimethacrylates and their hydrogels, Lin-Gibson, S., Bencherif, S., Cooper, J.A., Wetzel S.J., Antonucci, J.M., Vogel, B.M., *et al.*, *Biomacromolecules* 5, 1280–1287, 2004.

7 Control of lipase digestibility of emulsified lipids by encapsulation within calcium alginate beads, Li, Y., Hu, M., Du, Y., Xiao, H., and McClements, D.J., *Food Hydrocolloids* 25 (1), 122–130, 2011.

8 Characteristics of polyion complexes of chitosan with sodium alginate and sodium polyacrylate, Takahashi, T., Takayama, K., Machida, Y, and Nagai, T., *International Journal of Pharmaceutics* 61 (1), 35–41, 1990.
9 Impact of encapsulation within hydrogel microspheres on lipid digestion: an in vitro study, Matalanis, A. and McClements D.J., *Food Biophysics* 7 (2), 145–154, 2012.
10 Control of β-carotene bioaccessibility using starch-based filled hydrogels, Mun, S., Kim, Y.-R., and McClements, D.J., *Food Chemistry* 173, 454–461, 2014.
11 Effect of an organized lipid matrix on lipid absorption and clinical outcomes in patients with cystic fibrosis, Lepage, G., Yesir, D., Ronco, R., Champagne, R., Bureau, N., Chemtob, S., *et al.*, *Journal of Pediatrics*, 141, 178–185, 2002.

Index

Polyurethane Immobilization of Cells and Biomolecules: Medical and Environmental Applications, First Edition. T. Thomson.